Understanding JCT Standard Building Contracts

Understanding JCT Standard Building Contracts

Sixth edition

David Chappell

London and New York

First published 1987 by International Thomson Publishing, reprinted 1988, 1989

Second edition published 1991 by E & FN Spon, an imprint of Chapman & Hall, reprinted 1992

Third edition published 1993, reprinted 1994

Fourth edition published 1995, reprinted 1996, 1997

Fifth edition published 1998, reprinted 1999 (twice)

This edition first published 2000
by E & FN Spon
11 New Fetter Lane, London EC4P 4EE

Simultaneously published in the USA and Canada
by E & FN Spon
29 West 35th Street, New York, NY 10001

Reprinted 2001, 2002 by Spon Press

Spon Press is an imprint of the Taylor & Francis Group

Typeset in Sabon by Taylor & Francis Books Ltd
Printed and bound in Great Britain by Biddles Ltd, Guildford and King's Lynn

British Library Cataloguing in Publication Data
A catalogue record for this book is available from the British Library

Library of Congress Cataloging in Publication Data
Chappell, David.
Understanding JCT Standard Building Contracts /
David Chappell – 6th edn
Includes bibliographical references and index.
1. Construction contracts – Great Britain. I. Title.
KD1641 .C488 2000
343.41′07869–dc21 00-33893

ISBN 0–415–23107–8

Contents

Preface to the sixth edition

I continue to be surprised by the success of this little book among architects, quantity surveyors and contractors. I am stunned although gratified to learn that it has been adopted as a standard text for students in schools of architecture and building as well as being popular with those who are established in the industry. I will do my best to ensure that the text remains relatively simple and easy to read, and free from legalisms. The original intention was to provide a straightforward guide to the three standard forms of contract in common use.

My guiding principle has been the kind of book I would have wanted when I was a newly qualified architect. What I wanted then and what was not available was a short book which told me all I needed to know about the then current forms of contract and which I could read without having to look up every other word in a legal dictionary. I wanted a book which was not too superficial, which gave me a few insights and which pointed the way to further reading. With every revision comes the temptation to elaborate. I have kept my original intention in mind. Although there are many changes, the book should not be significantly longer than the last edition.

This edition has been thoroughly updated to take account of legal decisions and the Joint Contracts Tribunal have issued revised versions of JCT 80, IFC 84 and MW 80; these are now JCT 98, IFC 98 and MW 98 respectively. The new contract versions have been amended to take into account the Housing Grants, Construction and Regeneration Act 1996 (in Northern Ireland the Construction Contracts (Northern Ireland) Order 1997) and certain recommendations of the Latham Report. These changes are very significant and the indications are that the construction industry as a whole has not yet got to grips with them. An attempt has been made to explain the implications in simple terms.

As always, I am grateful to all those who have taken the trouble to express a view on this book and all suggestions have been carefully considered and acted upon where appropriate.

My customary thanks are due to my wife Margaret who has shown her usual patience throughout.

Note: The male pronoun is used in this book. This is for ease of reading and should be taken to mean both female and male individuals.

David Chappell
c/o 27 Westgate
Tadcaster
North Yorkshire
LS24 9JB

April 2000

List of abbreviations used in references

AC	Appeal Cases
ALJR	Australian Law Journal Reports
All ER	All England Law Reports
BCL	Building and Construction Law
BLM	Building Law Monthly
BLR	Building Law Reports
Ch App	Chancery Division Appeal Cases
CILL	Construction Industry Law Letter
CLD	Construction Law Digest
CLJ	Commonwealth Law Journal
Con LR	Construction Law Reports
Const LJ	Construction Law Journal
DLR	Dominion Law Reports
EG, EGCS	Estates Gazette Cases
LGR	Local Government Reports
LT	Law Times Reports
QB	Queen's Bench Law Reports
QBD	Law Reports, Queen's Bench Division
WLR	Weekly Law Reports

Introduction

This book is written as a helpful general guide to three popular forms of contract in the JCT series, namely: The Standard Form of Building Contract With Quantities, 1998 (JCT 98), The Intermediate Form of Building Contract, 1998 (IFC 98) and the Agreement for Minor Building Works, 1998 (MW 98). The text refers to contracts in England and Wales and substantially to Northern Ireland. It does not apply to Scotland which has major legal and contractual differences.

It has been thought sensible to arrange the guide under a series of topics rather than undertake a clause-by-clause interpretation, because there are dangers in looking at the clauses in isolation and taking too literal an approach. References to case law have been inserted for the benefit of those who wish to read further and to show the way in which contract provisions have been interpreted by the courts. It should be noted, however, that detailed expositions of cases have been excluded. The book tries to state the law and the position under the contracts as at the end of March 2000.

Legal language has been avoided and, where the contractual position is obscure, a suggested course of action is laid down. In the interests of clarity, the provisions have been simplified; this book, therefore, supplements but does not take the place of the original forms. More than anything, it is intended to be practical with emphasis on the contractor's interests. When in difficulty, the golden rule is to obtain expert advice.

1 Contractor's obligations

1.1 The forms

It seems appropriate to begin by looking briefly at the standard forms under consideration. All the JCT forms of contract were substantially amended in April 1998 to take account of the Housing Grants, Construction and Regeneration Act 1996 and the Latham Report. All the forms were reprinted at the end of 1998. JCT 80, IFC 84 and MW 80 became JCT 98, IFC 98 and MW 98 respectively.

JCT 98 is a very comprehensive document which is suitable for use with any size of building works. Due to its complexity, however, its use is likely to be reserved for projects which are substantial in value or complex in nature. There was initially some resistance to this form by architects and employers alike on its introduction in 1980. This resistance has now disappeared; very rarely contracts are let on the basis of its predecessor JCT 63.

There are dangers to the employer in using JCT 63:

- Unlike the other current JCT forms, it is no longer considered to be a negotiated contract. It falls within section 3 of the Unfair Contract Terms Act 1977 as the employer's 'written standard terms of business'. This means that no term in the contract will be taken to exclude the employer's liability if he is in breach and no term will be interpreted so as to entitle the employer to do less than is reasonably expected of him unless such terms satisfy the stringent tests of reasonableness. In simple terms the employer will often be prevented from wriggling out of a sticky situation simply by pointing to a contract clause.
- Anything in the form which is ambiguous or difficult to interpret will be construed by the courts against the employer (the *contra proferentem* rule).
- Fluctuation provisions are not 'frozen' if the contractor fails to complete by the completion date.
- There is no provision to defer possession of site. Any failure to give possession is, therefore, a serious breach of contract.

- Restrictions on order of work, hours of work and the like cannot be enforced.
- The determination provisions are weighted against the employer.
- The extension of time and the loss and/or expense clauses are poor.
- The nomination provisions are defective.

IFC 84 (now IFC 98) was introduced to fill the gap between JCT 80 (now JCT 98) and MW 80 (MW 98). Practice Note 20 (revised August 1993) issued by the JCT suggests its use if the Works (all the work to be done) are of simple content, adequately specified or billed and without complicated specialist work. The suggested upper price limit is £280,000 (at 1992 prices) and the maximum contract period of 12 months. Price and length of contract period are not, however, the most important factors. IFC 98 is only slightly shorter than JCT 98, but the layout is better and, although complex, it is relatively easy to read. A more user-friendly layout is being developed for JCT 98.

MW 98 is recommended for use on projects having a maximum value of £70,000 (at 1992 prices). It is not suitable for complex Works and no provision is made for bills of quantities or nominated sub-contractors. Very importantly, as far as contractors are concerned, there is only limited provision for reimbursement of loss and/or expense, although a claim can always be made using common law rights. This form is very popular and not only for minor Works. It is known for it to be used in conjunction with bills of quantities, although quite unsuitable. The reason for its popularity is no doubt because it is short and simply expressed. Its simplicity is deceptive, however, and there are pitfalls for the unwary.

The contractor may think that the suitability or otherwise of a particular form for a particular project is academic in the sense that he can do very little about it. The choice is for the employer advised by the architect. A thorough knowledge of the contents of the various forms, however, can influence the contractor's tender – if he has any sense.

Some employers use the standard forms, but with amendments to suit their own requirements and ideas. Such amendments, if substantial, may turn a standard form into the employer's 'written standard terms of business' under section 3 of the Unfair Contract Terms Act 1977 with the result referred to earlier. Amended forms of contract do have an unfortunate habit of backfiring on the party, making the amendments inconsistent or inoperative [1]. Any amendment to clause 25 of JCT 98 is likely to provide a bonus to the contractor unless great care is taken. More will be said about this later on when dealing with extensions of time.

Whichever of these forms is used, the contractor undertakes to carry out the Works in accordance with the contract documents.

It should be noted that the 1998 reprints of the JCT forms are not the same as the previous issue plus JCT Amendments. For example, IFC 98 is not the same as IFC 84 plus JCT Amendments 1–12. JCT have taken the opportunity to carry out many corrections. The intermediate form is particularly affected, because Amendment 12 was issued containing errors. If tenders have been invited using IFC 84 plus Amendments and the contract has been drawn up on the basis of IFC 98, the plain fact is that the formal contract will not properly reflect the agreement. Whether that is to the advantage or disadvantage of one or other party will depend upon particular circumstances.

Contract documents

It is vitally important to know which are the contract documents, because they are the only documents which spell out what the employer and the contractor have agreed to do. Letters exchanged before the contract is entered into and the contractor's programme are not contract documents, i.e. they are not binding on the parties, unless expressly so stated. Architects may point to minutes of site meetings as evidence of what was agreed, but they cannot amend the contract documents [2]. In order to amend the terms of the contract it would be necessary for the employer (not the architect on the employer's behalf) and the contractor formally to agree the change, preferably in writing and preferably as a deed. JCT 98 defines them in clause 1.3 as the contract drawings, contract bills, articles of agreement, conditions and appendix. The last three are contained in the printed standard form. The contract drawings must be the drawings on which the contractor tendered. It is not unusual for the architect to have made revisions to the original drawings between tender and the signing of the contract. The contract drawings must be carefully scrutinised before signing and, if such revisions are present, the architect must be asked to restore them to their previous condition.

IFC 98 provides four options:

- contract drawings and specification priced by the contractor;
- contract drawings and schedules of work priced by the contractor;
- contract drawings and bills of quantities priced by the contractor;
- contract drawings and the sum the contractor requires for carrying out the Works;

together with the agreement and conditions (the printed form) and, if applicable, particulars of tender of any named person in the form of tender and agreement NAM/T.

If the contractor is simply asked to state a sum required to carry out the Works – the last option – he must also supply a contract sum analysis or a schedule of rates on which the contract sum is based.

Strangely, neither the contract sum analysis nor the schedule of rates is a contract document. This may be important because one of these two documents is essential to value architects' instructions requiring a variation (clause 3.7.1). It will normally be to the contractor's advantage if the third option is used (with priced bills) because it puts the onus on the employer to ensure that the quantities are correct. All the other options provide room for dispute if there are inconsistencies.

MW 98 provides, in the first recital, for the contract documents to be any combination of contract drawings, specification and schedules together with the conditions (the printed form). The second recital provides that the contractor must price either the specification or the schedules or provide a schedule of rates. In the latter case, the schedule of rates would not be a contract document.

It is usual to talk about 'signing' the contract, but in fact it can be executed in either of two ways: under hand (also known as a 'simple' contract) or as a deed (also known as a 'specialty' contract). As far as building contracts are concerned, there are two important differences. A deed does not need what is called 'consideration' to make it a valid contract, but a simple contract does need consideration. For example, a builder who offered to construct a house for someone would have to receive something in exchange for there to be a valid simple contract, but if the contract was entered into as a deed, it would be binding even if the builder agreed to build the house without any reward.

The second important difference concerns the Limitation Act 1980 which operates to limit the period during which either party may bring an action to enforce their rights under the contract. In the case of a simple contract, the period is six years from the date of the breach. In the case of a deed, the period is 12 years. It is clear, therefore, that a contractor is more exposed if he enters into a building contract in the form of a deed.

Until quite recently, a deed had to be sealed in order to be properly executed. This was usually achieved by the impression of a device on wax or a wafer and fixed to the document. In fact, it was usually sufficient if it could be shown that both parties intended the document to be sealed [3]. The Law of Property (Miscellaneous Provisions) Act 1989 and the Companies Act 1989 have abolished the necessity for sealing for individuals and companies respectively. Indeed, sealing alone is not sufficient to create a deed (except in Northern Ireland where the Acts do not apply). All that is necessary now in order that a document be executed as a deed is that it be made clear on its face that it is a deed and, in the case of a company, that it is signed by two directors or a director and company secretary, or, in the case of an individual generally, that it is signed by the individual in the presence of a witness who must attest the signature. It is a matter on which proper advice should be sought before executing the contract.

Clause 1.11 of JCT 98 and clause 1.16 of IFC 98 allow the parties to enter into an agreement that communications can be exchanged electronically, i.e. by email. Security may be a factor to consider and the use of an electronic signature is sensible. The details of such an agreement are provided in annex 2 to the conditions. It should be noted, however, that there are certain notices under the contract which must be served in writing in the ordinary way. These are:

- determination of the contractor's employment;
- suspension by the contractor of the performance of his obligations;
- the final certificate;
- any notice given to invoke the dispute resolution procedures;
- any agreement the parties make to amend the contract or the electronic data interchange provisions.

It appears, therefore, that an interim certificate could be issued by electronic data interchange if the parties enter into the appropriate agreement.

Discrepancies

It is quite usual for there to be some small, and sometimes large, discrepancies between the provisions in the printed form and in, say, the bills of quantities or specification. Priority of documents then becomes important. It is often thought that terms which are hand- or type-written must take precedence over those which are printed because the written terms must represent the clear intentions of the parties. Indeed, that is the general law: type prevails over print [4]. However, JCT 98 clause 2.2.1, IFC 98 clause 1.3 and MW 98 clause 4.1 clearly stated that nothing contained in any of the contract documents will override or modify the terms in the printed form.

This kind of provision has been upheld in the courts [5]. In practice, it means that, if a term in the contract bills or specification is in conflict with a term in the printed form, the printed term will prevail. For example, if the bills provide for an estate of houses to be completed on specific dates, in other words phased completions, and the printed form contains just one date, it is the date in the printed form which will apply and the contractor will have fulfilled his obligations as to the time for completion if he completes all the houses on that one date. If the printed form stipulates that the employer must honour certificates within 14 days, that stipulation cannot be overridden by a clause in the bills allowing 21 days. To be effective, the change must be made to the printed form itself.

All three contract forms are lump sum contracts. That is to say that, in general, the contractor takes the risk that the work may be more costly than he expects. Specific clauses, however, modify the effects.

In particular, when bills of quantities are used, JCT 98 clause 2.2.2 and IFC 98 clause 1.5 expressly provide that the bills are to be prepared in accordance with the Standard Method of Measurement (SMM) unless specifically stated otherwise in respect of particular items. Errors in the bills are to be corrected and treated as variations. The effect of this is that a contractor is entitled to price everything as though measured in accordance with SMM unless there is a note for that item. A general note to the effect that not everything is measured in accordance with SMM would not be effective. Contractors can recover substantial amounts of money simply by paying attention to this point.

If the contractor makes an error in pricing which is not detected before acceptance of the tender, he is stuck with it. This may seem harsh and many *ex gratia* claims are made on this basis, but if two parties contract together they are usually taken to know what they are doing. It may be possible for the contractor to obtain relief if he can show that the employer detected a substantial mistake in pricing and, knowing that the contractor would not wish to contract on those terms, purported to accept the tender [6]. In Canada, an architect has been held liable to a contractor for failure to include important information in the invitation to tender. As a result, the contractor was unable to use his preferred system and lost money [7].

It is now established that the contractor has no duty to search for discrepancies and inconsistencies between the contract documents. The architect is responsible for providing the contractor with correct information. If the contractor does not discover a discrepancy until too late, the employer must pay for any additional costs resulting [8]. In the nature of things, inconsistencies will be present. All the forms make provision for the correction of such inconsistencies: JCT 98 in clause 2.3, IFC 98 in clause 1.4 and MW 98 in clause 4.1.

The architect may well consider that a contractor who is carrying out his obligations properly will have to examine the documents carefully and so detect discrepancies. This favourite argument does not change the legal position noted above. Of course, upon finding a discrepancy, the contractor should always ask the architect, in writing, for an instruction.

Most of the contractor's problems in this respect arise because he is anxious to proceed 'regularly and diligently' and when confronted by two drawings, or a drawing and a bill item, which do not correspond, he attempts to solve the problem himself. By so doing, the contractor loses his right to payment for any variation and probably takes responsibility for the design of that particular part of the work. If, however, the contractor does not proceed on that particular part of the work, asks the architect for instructions and notifies delay and disruption, he will be entitled to the whole of his loss, in terms of time and money as well as payment for the variation instruction when it eventually arrives.

A closely linked situation, although not strictly an inconsistency, is where the contractor is not provided with a particular detail he requires. He may think he knows what the architect intends to be done, but again he should beware of doing it without precise instructions. He should deal with it in precisely the same way as inconsistencies.

Although payment for architect's instructions requiring a variation to correct a discrepancy will be fairly automatic under the terms of JCT 98, the same may not be true of IFC 98 where bills are not used and MW 98.

For example, under the provisions of clause 1.2 of IFC 98, the contractor will be deemed to have priced for work in the specification even if not quantified. The rules in this clause are quite complicated and will repay careful study before tendering. Broadly, the rule is that if there are no quantities for a particular item, the contract documents must be read together. If there is a conflict between documents, the drawings prevail. Where quantities are shown, they prevail.

One of the dangers of the deceptively simple MW 98 is that the contractor's obligation is to carry out the Works in accordance with the contract documents on which he has submitted his price. The contract documents will probably be drawings and specifications and, therefore, he will be deemed to have included for the whole of the work shown on the drawings and specification. If there is an inconsistency, it appears that the contractor must be deemed to have priced for either option. In consequence, it appears that cases where the contractor will be entitled to additional payment as a result of inconsistencies will be rare.

1.2 Implied and express terms

The general law will imply three important terms into all building contracts:

- The contractor will carry out his work in a workman-like manner.
- The contractor will use and supply good materials.
- The contractor will undertake that the completed building is reasonably fit for its intended purpose where that purpose is known and there is no other designer involved.

These terms can only be modified or supplanted if there are express terms (i.e. written in) in the contract dealing with the same topics or, in the case of the third term, if someone has been employed in a design capacity.

Implied terms come as a shock to some contractors, who think that the whole of their obligations are covered by the terms which can be read in the printed document. There are, in fact, a good many other terms which the law will imply.

Some of these terms are implied by statute, for example, the Supply of Goods and Services Act 1982 and the Defective Premises Act 1972. In addition, the Unfair Contract Terms Act 1977 will often operate to void the effect of some clauses in a contract, particularly those clauses seeking to exclude or reduce liability.

The courts are usually reluctant to imply terms into a contract made between two parties because, provided that there is no mistake, misrepresentation, illegality, etc. in a contract, it is generally considered that the parties should be left with the bargain they have made, no matter how uncomfortable it turns out to be. When the courts do imply a term into a contract, it must not be inconsistent with an express term and it must have as its basis the presumed intentions of the parties; in other words, what they would have written into the contract had they given thought to the matter before the contract was made. The courts, however, will not imply terms into a contract simply because it seems to be a good idea. Terms will only be implied in accordance with certain principles which the courts themselves have built up over a number of cases. In general, the courts will imply a term if:

- It is the kind of term which, if asked, both parties would immediately agree was part of their bargain.
- The contract is apparently complete, but lacks just one term to make it workable.
- The parties have not fully stated the terms and the court is seeking to define the contract.
- In all the circumstances, it is reasonable to do so (rarely).

It can readily be seen that these categories are not completely separate, but rather instances of differing emphasis [9].

In JCT 98, IFC 98 and MW 98 the contractor's obligations are stated in clauses 2.1, 1.1 and 1.1 respectively. It is no accident that these obligations are at the very beginning of the conditions. In each contract the obligations are expressed in very similar terms: the contractor is to carry out and complete the Works in accordance with the contract documents. This is a basic and absolute obligation from which the contractor can only expect relief in very restricted circumstances; for example, if the employer prevents completion [10], or if the contractor lawfully determines his own employment in accordance with the appropriate clause (JCT 98 clause 28, IFC 98 clause 7.9 or 7.10, MW 98 clause 7.3). It is considered that this clause is not sufficient to impose any design liability on the contractor [11].

Contractor's duties

The contractor's duties are scattered throughout each contract. There are over 90 separate instances in JCT 98, more than 50 in IFC 98 and 15 in MW 98. These, it must be remembered, are express duties and take no account of the duties which will be implied. It is, therefore, of the utmost importance that the contractor reads the contract carefully. It should be considered a working tool, not something to be thrown into a drawer. It is well worthwhile highlighting the duties in coloured pen.

Very often, the contractor's entitlement to money or extension of time depends upon the correct performance of some obligation as a precondition. A good example is to be found in JCT 98 clause 26.1.1, where a precondition to the ascertainment of loss and/or expense is that the contractor must make application as soon as it has become, or should reasonably have become, apparent that the regular progress of the Works has been or is likely to be affected. Failure to carry out the duty precisely could seriously prejudice the contractor's chances of obtaining reimbursement. Thus, a contractor who waits until several weeks after the event might well be considered to be in breach of his duty.

An obligation which seems to cause more difficulty is contained in JCT 98 clause 23.1 and IFC 98 clause 2.1. This is the obligation to proceed with the Works regularly and diligently. Clause 1.1 of MW 98 refers to carrying out the Works with due diligence but it is thought that they have much the same effect. The question is, what do they mean?

The words will be interpreted in the light of the facts, but it seems clear that simply going slow will not necessarily be taken to mean that the contractor is not proceeding diligently. It has been suggested that an obligation to work with due diligence is an obligation on a contractor to work so as to meet the key dates and the completion date in the contract [12]. It has been established that the duty is to proceed regularly and diligently (the words being read together), efficiently and industriously, using the resources necessary to complete the contract on time. It is not sufficient for him to have men on site at regular intervals; the work must be being progressed [13]. This is important in view of the fact that failure to proceed regularly and/or diligently is a ground for determination by the employer in all three contracts. Failure on the part of the contractor to keep to his programme would hardly qualify under this head unless such failure was gross. But an architect who did not issue a default notice in the face of the contractor's clear failure to proceed regularly and diligently would be in breach of his duty to the employer and the court of appeal has recently explained the meaning of regularly and diligently as follows:

> What particularly is supplied by the word 'regularly' is not least a requirement to attend for work on a regular daily basis with sufficient in the way of men, materials and plant to have the physical capacity to progress the Works substantially in accordance with the contractual obligations.
>
> What in particular the word 'diligently' contributes to the concept is the need to apply that physical capacity industriously and efficiently towards the same end.
>
> Taken together the obligation upon the contractor is essentially to proceed continuously, industriously and efficiently with appropriate physical resources so as to progress the Works steadily towards completion substantially in accordance with the contractual requirements as to time, sequence and quality of work [14].

JCT 98 clause 5.3.1.2 requires the contractor to provide two copies of his master programme and to update it within 14 days of any decision by the architect in relation to extensions of time. IFC 98 and MW 98 have no such provision, but there is no reason why such a requirement should not be incorporated into the specification or preliminaries. It is useful if the programme is specified as being in network form, because it enables the estimation of extensions of time much more easily than with a simple bar chart. The courts have approved the use of computerised techniques to input the delays into a precedence diagram to determine the likely extension of time [15]. All architects and contractors should make use of this facility.

Contractors often show on their programmes that they intend to complete several weeks before the Date for Completion in the contract. It is the contractor's privilege to complete earlier than the Date for Completion if he wishes, but the employer, through his architect, is not bound to assist him by providing information earlier than necessary to enable the contractor to complete the Works by the contractual date [16]. It may not be to the contractor's advantage to produce a programme showing that he will finish four weeks early. The architect may argue that the contractor is not entitled to any extension of time until delays affect the end date by more than four weeks. Of course, in such an instance the contractor would be entitled to recover whatever direct loss and/or expense he could prove if matters giving rise to loss and/or expense were the cause of the delay.

Architect's satisfaction

The contractor's basic obligation is qualified in each of the forms under consideration by the proviso that where and to the extent that the approval of quality of materials or standards of workmanship are a matter for the opinion of the architect, they must be to his reasonable satisfaction. This does not impose a dual responsibility upon the contractor unless the contract documents expressly state that is to be the situation [17].

In general, the contractor must follow the requirements laid down in the contract documents, such as to provide timber to a certain standard. Provided the contractor does just that, he has fulfilled his obligations. It matters not that the architect is not satisfied, provided the materials or workmanship are in accordance with what is laid down in the contract. The architect may not be satisfied simply because he is disappointed with the results of his specification, but if he wants an improved result in such a case, the employer has to pay for it.

If, however, materials or workmanship is stated in the contract documents to be to the architect's satisfaction, or some such similar phrase, the contractor's obligations will be to satisfy reasonably the architect. The architect cannot specify iron and expect gold. In effect, the architect's approval will override any specification requirement. It is, therefore, important to obtain such approval in writing.

Even if the contractor does not manage to ensure that the architect expresses his approval in writing on every occasion he expresses it orally, the issue of the final certificate, in the case of JCT 98 and IFC 98, will be conclusive evidence that such approval has been given (JCT 98 clause 30.9.1.1 and IFC 98 clause 4.7).

This is of crucial importance, especially to those contractors who may be worried at tender stage about the frequency of architect's approvals in the specification. Of course, it does not remove the obligation to satisfy the architect, but once satisfied and the final certificate issued, it becomes the architect's worry.

In 1994 it was established that quality and standards of materials and workmanship are always matters for the opinion of the architect [18]. The effect was that the final certificate under the then JCT 80 or IFC 84 was conclusive that the architect was satisfied with these matters whether or not they had been mentioned in the specification or bills of quantities. Clause 30.9.1.1 of JCT 98 and clause 4.7 of IFC 98 have been amended by the JCT with the object of restoring the situation to the pre-1994 position. The Royal Institute of British Architects (RIBA) produced a series of stickers (A, B and C) for architects' use, together with a form of Declaration to be signed by the contracting parties. Few contractors agreed to sign (wisely). Even if signed it is doubtful whether it was effective unless executed as a deed. The stickers are unlikely to have been effective. In the case of MW 98, the final certificate is not conclusive about any aspect of the work.

If the architect does not issue the final certificate at the correct date, the employer will not be able to take advantage of the failure in order to take action against the contractor for defects if the action would have been prevented if the final certificate had been correctly issued [19].

JCT 98 clause 8.2.2 stipulates that where materials, goods or workmanship are to be to the architect's satisfaction, he must express any dissatisfaction within a reasonable time of the work being carried out. In the light of the remarks above, it seems that clause 8.2.2 extends to all materials, goods and workmanship whether or not expressly noted in the specification or bills of quantities. This is an added safeguard for the contractor against the architect waiting until the project is nearly complete before making known his complaint. It puts the onus on the architect to decide whether or not he is satisfied. The precise meaning of 'a reasonable time' will always be open to dispute, but in this context it is thought that the architect must express his dissatisfaction before the contractor carries out the next operation. For example, if the contractor lays a screed, the architect must say if he is not happy with it before the floor finish is applied and possibly before all wet trades have finished.

Design

In general, the contractor will have no obligation to, nor liability for, design unless he has taken it upon himself (not uncommon), or unless the contract documents clearly set out such an obligation and liability. That position is modified by clause 42 of JCT 98. It was always possible to introduce the contractor's design liability into JCT 80 by adding the Contractor's Designed Portion Supplement. Clause 42 is intended to achieve the same result on a smaller scale. Architects often specify items which unavoidably involve design and the contractor secures the items by means of sub-contractors or suppliers. Problems arise when design defects appear. The architect should use Clause 42 when he requires the supply of such things as roof trusses, precast floor beams or any other element having a design content and which the architect does not wish to design himself or through a consultant. It is referred to as 'performance specified work' and it must be identified in the Appendix, shown on the contract drawings and a performance specification must be included in the bills of quantities.

The clause is lengthy. There must either be enough information for the contractor to price the work or a provisional sum must be included. After the contract is entered into, the architect may not instruct the contractor to carry out performance specified work which has not been included in the bills in one form or another. The contractor has to give the architect a statement describing how he intends to carry out the work. By written notice, the architect can require the contractor to correct any deficiencies in the statement. The following points should be noted:

- Correction of errors in the information which must be included in the contract bills is to be treated as variation required by architect's instruction.
- The architect is responsible for integrating the performance specified work into the rest of the design.
- The contractor may object to an instruction which he considers will badly affect the performance specified work.
- The contractor's obligation is to use reasonable skill and care.
- Performance specified work is not to be provided by nominated sub-contractors or suppliers.

Duty to warn

Although the position is not crystal clear, it seems that the contractor has no duty to warn the architect of defects in his design although, in certain circumstances, he may have a duty to warn the employer if it can be shown that the employer is placing special reliance upon him [20]. He certainly has a duty to the employer in those cases where the architect has produced the original drawings, but is taking no further part in the project

[21]. The prudent contractor will always notify the architect if he considers that there is a design defect irrespective of whether he has a duty to do so [22]. It is wise to put the notice in writing. If the architect ignores it, he does so at his peril, but then no blame can attach to the contractor.

1.3 Materials and workmanship

JCT 98 clause 2.1, IFC 98 clause 1.1 and MW 98 clause 1.1 impose substantial obligations on the contractor to carry out and complete the Works in accordance with the contract documents. Moreover, items for the architect's approval, so far as quality and standards are concerned, are to be to his reasonable satisfaction. These obligations form a basis against which the contract clauses dealing with materials and workmanship should be studied.

MW 98 makes no further reference to materials, goods and workmanship except that materials and goods reasonably and properly brought onto site and adequately stored and protected are to be included in progress payments (clause 4.2).

Procurable

A very useful protection is given to the contractor by JCT 98 clause 8.1. His duty is stated to be to provide materials and goods of the standards described 'so far as procurable'. That is, obtainable. If the contractor cannot obtain goods, etc. of the standards described, his obligation seems to be at an end.

The clause provides no escape for a contractor who is finding it more difficult or more expensive to obtain the required standard of goods. Unexpected rises in prices are the contractor's risk. If the contractor undertook, at tender stage, to supply materials which were, even then, unobtainable or if the materials, etc. are unobtainable because he was late in placing his order, it is probable that he has a duty to supply materials, etc. at least equal in standards to what is specified and at no extra cost to the employer.

If the materials become unobtainable after his tender is accepted and no blame attaches to the contractor, it then falls to the architect to issue an instruction to vary the materials. If the new materials prove more costly, the contractor is entitled to be paid the difference. Neither IFC 98 nor MW 98 provide this relief for the contractor, who will be responsible under these contracts for providing alternatives at the original cost.

Workmanship has now been separated from materials and goods under JCT 98. So far as workmanship is concerned, it must be of the standard described in the bills of quantities (or specification if the Without Quantities version is being used). If there is no description, the standard must be appropriate to the Works. Such a term would be implied as a matter of law, but there is much room for argument and it is to be preferred if the workmanship is properly described. Both workmanship and materials must be to the reasonable satisfaction of the architect if he so requires in accordance with clause 2.1.

Clause 8.1.3 provides that all work must be carried out in a workman-like manner. If the contractor fails to comply with this provision, the architect may issue whatever instructions may be reasonably necessary, including the instructing of a variation, in consequence. The architect must first consult the contractor who must, if applicable, consult the relevant nominated sub-contractor. This is providing, and to the extent that they are necessary instructions, the contractor is not entitled to any payment for the variation nor for loss and/or expense, nor is he entitled to any extension of time. This clause is something of a 'catch all' attempt to cover those circumstances where the architect cannot say that the work is not in accordance with the contract, but it is not carried out in a workman-like manner so that it is dangerous or it threatens the proper carrying out of other parts of the Works.

Opening-up and testing

The architect is entitled to require the contractor to open up work already covered up or to carry out testing of materials, even if they are already built into the Works (JCT 98 clause 8.3 and IFC 98 clause 3.12). MW 98 makes no specific reference to opening-up and testing, but the architect probably has the power to order such work under clause 3.5. The contractor is entitled to be paid for opening up and making good again and for any tests required unless:

- The cost of such opening-up and testing is already included in the contract.
- The work or materials is found to be not in accordance with the contract.

If it is suggested that the cost is already included in the contract, the contractor should make sure that it is indeed included. It seems unlikely that any very general note in the specification would cover the situation and the amount of opening-up and testing would have to be specified in reasonable detail. Otherwise, the contractor could be required to undertake limitless amounts of investigative work of this nature with no recompense. If uncovered work is found to be defective, the contractor has to bear the cost of uncovering, correcting the defects and making good again. That appears to be reasonable.

Contractors often feel aggrieved when asked to open up work. The grumble seems to be threefold: it throws doubt on their competence, the architect or clerk of works could have inspected before covering up and the contractor feels that, after opening-up, some excuse will be found to justify the exercise and leave the contractor with the expense. Thus, the architect who instructs opening-up in order to inspect part of the foundations, just before practical completion of the whole building is due, will not be a popular fellow.

Contractors just have to live with the fact that architects do not always trust them. Opening-up is often ordered because some aspect of the work causes the architect to suspect that all is not well. Take, for example, the case of a specified 60 mm concrete screed laid on a solid concrete ground floor with a separating membrane between. The architect may notice severe cracking, suspect that the screed is not thick enough and instruct opening-up of part of it to make sure. He may also order testing of pieces of the removed screed to check that it is of the correct mix, etc. If it is found that the screed is at least 60 mm thick and of the correct mix, etc. and the separating membrane is in position and undamaged, the contractor would have a good case for reimbursement.

Undoubtedly, the screed is defective in the generally accepted sense because it is badly cracked, but it is demonstrably in accordance with the contract, which is all the opening-up clause requires for the contractor to obtain payment. The problem may be that, in the circumstances, the architect should have specified a reinforced screed. On the other hand, if everything is found to be in accordance with the contract except that the screed is only 55 mm thick, it will avail the contractor nothing to protest that the 5 mm difference cannot possibly have caused the cracking. The contractor may argue that to avoid cracking, some light reinforcement would be necessary, that cracking, in itself, is not serious and so on, but he would still be expected to remove the screed and relay it to the required thickness at no extra cost. If, however, in the process of relaying, the architect instructed the inclusion of additional light reinforcement in the new screed, the contractor would have an arguable case that he should be paid for more than the additional reinforcement.

It should be noted that the architect has no duty to the contractor to find defects [23]. His duty is to the employer. The contractor can have no recourse against the architect who does not detect a defect until late in the contract, because it is the contractor's obligation to build in accordance with the contract documents.

Failure of work

An important provision in IFC 98 (clause 3.13) relates to failure of work. If any materials or work are found to be at variance with contract requirements, the contractor must, without prompting, tell the architect how he intends to ensure that there are no similar failures elsewhere on the job.

The proposed measures must be entirely at the contractor's own cost. For example, if it is discovered that the dpc has been omitted over some windows, the contractor might propose opening-up, say, 10 per cent of all window heads for inspection by the architect. If the contractor's proposal is accepted, the contractor has to stand the cost even though all inspected window heads have dpcs in accordance with the contract. He would, however, be entitled to an appropriate extension of time. The contractor must submit his proposal within seven days of the discovery of the initial defect.

If the architect is not satisfied with the contractor's proposal, if the proposal is not submitted within seven days or if there are some pressing safety or statutory reasons why he cannot wait even seven days, he may issue instructions himself for opening-up as he sees fit at the contractor's expense. In such a situation, the contractor has four days in which to decide whether to carry out the instruction or to write to the architect with his objections. Objections will usually relate to the amount of opening-up required but there could be other grounds. If the architect does not withdraw or modify his instruction within a further seven days, the matter is automatically referred to arbitration.

A problem for the contractor lies in the fact that, in this clause, no distinction is made between major and minor failures of work. The only qualification is that the proposals are restricted to establishing that there are no similar failures. On every contract there will be a multitude of very minor instances, readily corrected, where work is not in accordance with the contract and one or two cases which may be major. The contractor would be well advised to settle with the architect, at an early stage, in what circumstances he expects clause 3.13 to be operated. This is best done at the first site meeting and recorded in the minutes, or, less easily, by an exchange of letters.

JCT 98 clause 8.4 and IFC 98 clause 3.14 empower the architect to order the removal from site of work or materials which are not in accordance with the contract. MW 98 has no similar provision, but it is thought that the architect must have this power under clause 3.5. Note that an instruction to correct defective work is not valid (except possibly under MW 98 terms). The instruction must require removal of the work from the site [24]. In practice, of course, it usually amounts to the same thing.

Under JCT 98 if materials, goods or workmanship are not in accordance with the contract, the architect may do any or all of the following:

- instruct that the materials, etc. be removed from site;
- allow the work to remain and make an appropriate deduction from the contract sum. The employer must agree and the architect must consult the contractor who, in turn, must consult any nominated sub-contractor who may be affected;
- issue such instructions requiring a variation as are reasonably necessary as a consequence of the architect's previous action under

this clause. There is to be no additional cost, extension of time or loss and/or expense. Once again, the architect must consult the contractor who must consult the nominated sub-contractor if affected;
- issue instructions under clause 8.3 requiring the contractor to open up or test the work to establish to the architect's reasonable satisfaction the likelihood of any other similar instances of failure to comply with the contract. If the instruction is reasonable, the contractor is not entitled to any addition to the contract sum, but he is entitled to an extension of time unless work is found not to comply with the contract.

In using the last power, the architect is required to have 'due regard' to a code of practice. The code is part of the printed contract form and it is a list of relevant factors the architect is required to consider in order that the extent of his instructions to open up is reasonable. There are 15 items ranging from the need to show the employer that a particular defect does not occur throughout the Works, to proposals the contractor may make and 'any other relevant matters' – which just about covers everything. The architect is not required to justify his instruction to the contractor. Reference to the code will probably be most useful if the contractor decides to seek arbitration on whether his objection to the instruction is justified. It should be noted that the architect's decision to issue instructions under this clause is not open to review by an arbitrator. It seems, however, that the correctness of such a decision can be the subject of arbitration.

There is a danger for the employer if the architect instructs how defective work is to be remedied. Some architects appear to think that because the contractor is in breach of contract in providing materials or workmanship which is not in accordance with the contract, they can order removal from the site and replacement with more expensive work or materials. If the architect does give such instructions, the contractor is entitled to be paid as though the architect were ordering a variation [25].

A question which sometimes arises is the extent to which a contractor is liable if the architect's specification allows the contractor to choose the precise type of material within limits. It is now established that, unless there is special clause in the contract which provides to the contrary, the architect is liable for the suitability of the materials he specifies. If he specifies in such a way that the contractor has a choice, it may be assumed that the architect believes that no further restrictions are needed. Provided the material chosen was good of its kind and he did not have actual knowledge of its likely bad effects, the contractor will not be liable [26].

Employer's representative

This provision only occurs in JCT 98. Clause 1.9 allows the employer to appoint a representative who can act for him under the contract and exercise all the functions which the contract allows the employer to perform, or prescribes that he shall do; in other words, the employer's powers and duties.

In order to achieve this, the employer must issue written notice to the contractor, stating that he wishes the representative to deal with all the functions of the employer under the contract or whether he requires exceptions. Any exceptions must be set out in the notice. For example, the employer may not want the employer's representative to make certain decisions.

An intriguing footnote ([p]) suggests that to avoid possible confusion, as it is put, over the quite distinct roles of the architect and the quantity surveyor and the role of the employer's representative, the employer's representative should not be either architect or quantity surveyor. Presumably, although not expressly stated, the employer's representative will be in reality a project manager in one of its several manifestations.

Ownership of goods

Much difficulty has been caused by what is known as 'retention of title'. The position can be quite complicated. Clauses 16 of JCT 98 and 1.11 of IFC 98 are intended to provide that materials and goods stored on or off site, for which the employer has paid, become his property.

In essence the problem is that the terms of the main contract only bind the parties to that contract; they cannot, subject to the Contracts (Rights of Third Parties) Act 1999, affect the rights of third parties such as suppliers. The supplier of the goods may have a retention of title clause in his contract of sale to the contractor which stipulates that the goods, in fact, remain the supplier's property until the supplier has been paid by the contractor. Therefore, if the employer pays the contractor for goods stored on or off site and the contractor does not pay the supplier, ownership of the goods will not pass to the employer. This is because the contractor cannot pass ownership until he has it himself. The goods still belong to the supplier who may, in the event of the contractor's liquidation, take them away.

In such a case, the employer may be faced with the prospect of paying twice for the same goods; hence the traditional reluctance on the part of the architect to certify payment for off-site materials (see chapter 5, for the revised provisions for off-site materials and goods). He is obliged to certify payment for materials stored on site, but in either case the contractor may have to produce evidence of ownership or satisfy the architect in some other way that the employer will become the owner of the goods when he pays the contractor.

A supplier's retention of title clause will normally be defeated once the goods are incorporated into the fabric of the building [27].

That is a very simple exposition of a very complex problem and contractors who encounter difficulties in this area should seek good legal advice. It tends, however, to be more of a problem for employers and suppliers.

It should be noted that all the contracts now contain a clause making it clear that no rights are conferred on third parties.

References

1. Bacal (Construction) Midlands Ltd v Northampton Development Corporation (1975) 8 BLR 88; Update Construction Pty Ltd v Rozelle Child Care Centre Ltd (1992) Building Law Monthly vol. 9.2.
2. James Miller & Partners v Whitworth Street Estates Ltd [1970] 1 All ER 796.
3. Whittal Builders Co Ltd v Chester le Street D C (1987) 11 Con LR 40.
4. The Brabant [1966] 1 All ER 961.
5. English Industrial Estates Corporation v George Wimpey & Co Ltd (1972) 7 BLR 122.
6. McMaster University v Wilcher Construction Ltd (1971) 22 DLR (3d) 9.
7. Auto Concrete Curb Ltd v South Nation River Conservation Authority and Others (1994) 10 Const LJ 39.
8. London Borough of Merton v Stanley Hugh Leach (1985) 32 BLR 51.
9. Liverpool City Council v Irwin [1976] 2 All ER 39.
10. Lawson v Wallasey Local Board (1982) 11 QBD 229.
11. John Mowlem & Co Ltd v British Insulated Callenders Pension Trust Ltd (1977) 3 Con LR 64.
12. Greater London Council v Cleveland Bridge and Engineering Co Ltd (1986) 8 Con LR 30.
13. ibid.
14. West Faulkner Associates v The London Borough of Newham (1995) 11 Const LJ 157.
15. John Barker Ltd v London Portman Hotels Ltd (1996) 12 Const LJ 277.
16. Glenlion Construction Ltd v The Guinness Trust (1987) 39 BLR 89.
17. National Coal Board v William Neal & Son (1984) 26 BLR 81.
18. Crown Estates Commissioners v John Mowlem and Company Limited (1994) 10 Const LJ 311.
19. Matthew Hall Ortech Ltd v Tarmac Roadstone Ltd (1997) 87 BLR 96.
20. University Court of the University of Glasgow v William Whitfield & John Laing (Construction) (1988) 42 BLR 66.
21. Brunswick Construction v Nowlan (1974) 21 BLR 27.
22. Edward Lindenberg v Joe Canning and Jerome Construction Ltd (1992) 62 BLR 147.
23. Oldschool & Another v Gleeson (Contractors) Ltd & Others (1976) 4 BLR 103.
24. Holland Hannen and Cubitts (Northern) Ltd v Welsh Health Technical Services Organisation (1981) 18 BLR 80.
25. Simplex Concrete Piles Ltd v Borough of St Pancras (1958) 14 BLR 80.
26. Rotherham Metropolitan Borough Council v Frank Haslam & Co Ltd and M J Gleeson (Northern) Ltd (1996) EGCS 59.
27. Reynolds v Ashby [1904] AC 406.

2 Insurance

2.1 General

Insurance is a highly specialised field. Although the details of policies are best left to the experts, it is essential that the employer and the contractor know their respective rights and obligations under the contractual provisions. JCT 98 deals with indemnities and insurance in clauses 20, 21 and 22. IFC 98 and MW 98 deal with them in clause 6 in each case. They can be broken down into the following parts:

- indemnities – injury to persons and property;
- insurance – injury to persons and property;
- insurance – liability of the employer (not MW 98);
- insurance of the Works – new Works – existing structures;
- insurance – liquidated damages (not MW 98);
- remedies if a party fails to insure.

JCT 98 and IFC 98 provisions are virtually identical. MW 98 provisions are less detailed with some significant omissions and they will be considered later.

2.2 Injury to persons and property

The contractor indemnifies the employer and takes liability in the case of any loss, expense, claim or proceedings as a result of carrying out the Works in respect of:

- personal injury or death of any person unless and to the extent that it is due to the act or neglect of the employer or anyone for whom he is responsible;
- injury or damage to property of all kinds except the Works which and to the extent that it is due to the negligence or default of the contractor, sub-contractors or agents.

The contractor is also liable if the injury to persons or property is a result of breach of statutory duty and, in addition to his servants or agents, he is made liable for 'any person employed or engaged upon or in connection with the Works' and any other person who may properly be on site in connection with the Works excluding those persons for whom the employer is responsible.

Thus, if a person is injured due to the Works, the contractor is liable unless it is the employer's fault. If property is damaged, the contractor is liable only if it is his fault [1]. It is clear that the contractor is still obliged to indemnify the employer against claims for personal injury or death even if the employer's neglect is only partially responsible. In such a case, of course, the contractor's liability would be reduced accordingly. A similar formula has been introduced to maintain the indemnity to the employer even if the contractor's negligence is only a part cause of injury to property.

In practice, a person suffering injury to his person or property would usually claim against the employer who would, by virtue of this clause, join the contractor as a third party in any proceedings. It should be noted that the contractor is not liable under this clause for any loss or damage to the Works unless they have been taken into partial possession under JCT 98 clause 18 or IFC 98 clause 2.11.

It may be thought superfluous to have an indemnity clause when the following clause requires the contractor to take out insurance against just the same liabilities. But if a claim was successful against the employer and the insurance company refused to meet the claim for some reason, the contractor would retain liability. The level of insurance required is to be inserted in the Appendix.

It should be noted that the contractor does not provide the employer with indemnity against the results of the employer's own negligence. To provide such indemnity, the clause must be expressed in very clear words [2].

2.3 Liability of employer

JCT 98 clause 21.1 and IFC 98 clause 6.2.4 are in identical terms intending to cover the liability of the employer if damage is caused to any property other than the Works by the carrying out of the Works. Such occurrences as collapse, subsidence, heave, vibration, weakening or removal of support or lowering of ground water arising from the carrying out of the Works are included. This clause operates when the damage which occurs is not due to negligence on the part of either employer or contractor or their servants or agents. For example, the contract may call for piling operations on a town centre site and, despite careful design and conscientious work, cracking occurs in adjacent buildings. Because the contractor is only liable for such damage caused by his negligence, without this clause, the employer would have to foot the bill [3]. The clause envisages that a sum may be included in the contract documents representing the indemnity required and the architect may instruct the contractor to take out appropriate insurance. The insurance must be taken out by the contractor in joint names with insurers approved by the employer, with whom policy and premium receipts must be deposited.

The contractor is expressly stated to have no liability in respect of injury or damage to any person, the Works, site or any property caused by ionising radiations or contamination by radioactivity from any nuclear fuel or from any nuclear waste from the combustion of nuclear fuels, radioactive toxic explosive or other hazardous properties of any explosive nuclear assembly or component, or by pressure waves caused by aircraft or other aerial devices travelling at sonic or supersonic speeds. These risks are called 'excepted risks'. Any such damage, therefore, will be the sole responsibility of the employer.

2.4 Insurance of the Works

There are two types of insurance risks in the contract.

- 'Specified perils' insurance – this is insurance previously known as 'clause 22 perils' (JCT 80) or 'clause 6.3 perils' (IFC 84). It provides for insurance against fire, lightning, explosion, etc.
- 'All risks insurance' – this is insurance against physical loss or damage to work executed and site materials, but *excluding* the cost of repairing, replacing or rectifying property which is defective due to wear and tear, obsolescence, deterioration, rust or mildew on any work executed or materials lost or damaged as a result of its own defect in design, plan, specification, materials or workmanship or any other work executed which is lost or damaged if such work relied for its support or stability on the defective work. Other exclusions include loss or damage arising from war, nationalisation or order of any government or local authority, disappearance or shortage only revealed on the making of an inventory and not traceable to any identifiable event, the exceptions with regard to ionising radiations, etc. previously excluded from clause 22 or 6.3 perils and certain other exclusions applicable only to Northern Ireland (civil commotion, unlawful and malicious acts, etc.). Therefore, risks such as impact, subsidence, theft and vandalism are included in this type of insurance.

Insurance of the Works falls into two categories: new Works and work to existing structures. There are three standard clauses: JCT 98, 22A, 22B and 22C (IFC 98, 6.3A, 6.3B and 6.3C). The first two clauses relate to new work where either the contractor or the employer, respectively, has the responsibility to insure. The type of insurance to be taken out is 'all risks'.

Not only the Works, but the value of any unfixed materials or goods delivered to the site must be insured including a percentage stated in the Appendix to cover professional fees. Full reinstatement value is required which will, of course, be greater than the simple value of everything on the site, but it does not include the additional costs of carrying out subsequent work later than intended nor the loss suffered by the employer due to delay. The insurance must be taken out in joint names with insurers approved by the employer with whom policy and premiums must be deposited. The contractor should obtain good advice. The risk does not include consequential loss [4].

The insurance must be maintained up to and including the date of the issue of the certificate of practical completion. This will almost always be some days after the actual date of practical completion and prevents the premature termination of insurance cover if, for some reason, expected practical completion does not take place. Provision is made for the cover to cease if determination of the contractor's employment occurs and this is the case even if either party contests the determination.

The term 'at the sole risk' was omitted from the 1986 amendments. The effect of this is that the insurers now have no right of subrogation against one or other of the parties if damage occurs due to the other party. Subrogation, in this context, is the right of the insurer to stand in the shoes of the insured party and recover, from the party causing the damage, amounts paid out. Since both parties are insured on a single policy, if one of the 'all risks' occurred due to the employer's negligence, while it is the contractor's obligation to insure in joint names, the position would be covered by the policy and the insurer would not be able to claim against the employer. This was the position under the old 1980 clause 22A, but not under clause 22B, leaving the contractor's position somewhat ambiguous. Now both clauses have been redrafted to give the parties equal protection. Where the contractor wishes to use his existing annual policy, it must contain terms which equate it to a joint names policy. It is no longer sufficient, or necessary, to have the employer's interest endorsed and the annual renewal date must be stated in the Appendix.

Financial responsibility

If there is any element of under-insurance or any excess payable on the insurance, this is the responsibility of the party who has the obligation to insure. Where the contractor has the obligation to insure, insurance money is to be payable to the employer, who must issue, through the architect, special 'reinstatement certificates' as repair work proceeds until the insurance money is fully certified. The contractor is entitled to no money additional to insurance money in respect of the insured matter, but he is entitled to interim certificates in the normal way for other work carried out even though some of that work has been damaged by an event which resulted in the insurance payment to the employer. Thus, except for the possibility of under-insurance or excess, neither the contractor nor the employer should suffer any financial loss. If the employer insures, the work is to be treated as if it were a variation and the contractor is to be paid its full cost.

If either the contractor or the employer insures, the contractor must immediately give written notice of any damage to employer and architect. He must state its nature, extent and location. After inspection by the insurers, if required, the contractor must proceed with restoration and removal of debris.

Existing structures

If the work consists of alterations or extensions to existing structures, the employer must insure the existing structure and contents, the Works and all unfixed materials delivered to the site and intended for incorporation. The insurance is in two parts:

- specified perils insurance for the existing building and contents;
- all risks insurance for the new work.

Both parts are to be taken out in joint names. The distinction between types of insurance is important. Where, for example, the contractor or his sub-contractor cause damage to a water pipe in an existing building, the contractor will be liable for the resultant damage caused by the escaping water which will not fall under the descriptions of 'flood, bursting or overflowing of water tanks, apparatus or pipes' as described in 'specified perils' [5].

In the event of loss or damage, the contractor must give immediate written notice to the employer and the architect. At this point, either party may determine the contractor's employment if it is just and equitable so to do (see section 7.2). If neither party opts for determination or the notice is decided against by an arbitrator, the contractor must make good the loss and damage, remove debris and proceed with carrying out the Works as before. The contractor's work is to be treated as a variation and he is to be paid accordingly.

It seems that fire due to a contractor's negligence is not covered by this insurance at least under MW 98 clause 6.3B [6].

Restoration

Note that the contractor's obligation to commence restoration under all the JCT 98 clause 22 (IFC 98 clause 6.3) insurances begins as soon as the insurers have carried out any inspections they deem necessary and not, as previously, when the claim is accepted. The contractor is, therefore, put into the position of carrying out rectification work before he knows whether any insurance money will be paid out.

Failure to insure

In regard to any of the insurance which is the contractor's responsibility, he is required to produce evidence of insurance on demand to the employer unless he has already deposited the policy and premium receipts. If he fails to insure, the employer may take out and maintain the appropriate insurance himself, deducting the cost from monies due or to become due to the contractor. Alternatively, he may sue for the debt.

The contractor has similar power of inspection if the employer fails to insure new work where that clause applied. If the employer fails to provide evidence of insurance, the contractor may himself take out the necessary insurance and he is entitled to have the amount of the premium added to the contract sum on production of a receipt. Where the employer is required to insure existing structures and extensions, etc. the contractor has unusual powers if the employer fails in his duty. The contractor has the right to enter and inspect the existing structure for the purpose of making an inventory and survey. Where a premium has been paid to take out the insurance, he is again entitled to have its cost added to the contract sum. The only difference between private and local authority use is that the local authority is not obliged to produce evidence of insurance on demand and the contractor has no power to take out insurance if the local authority fails to do so. These provisions, in any case, only apply if clause 22B or 22C (JCT 98) or 6.3B or 6.3C (IFC 98) is used.

Liability to insure

The revised wording probably overcomes a previous difficulty with this clause whereby, even if the contractor by his negligence caused fire damage to the Works, the employer had to bear the risk [7].

The contractor's liability to insure ends at the date of issue of the practical completion certificate, or on partial possession as regards the relevant part, but it is important that the contractor obtains formal certification before dispensing with such insurance [8].

Insurance of the contractor's plant and tools is not provided for under either contract and it is the contractor's liability. Most contractors do carry this sort of insurance.

If employer does not wish to insure

If the employer does not wish to use JCT 98 clause 22A, 22B or 22C, model clauses have been prepared which enable the employer to take either the risk or the sole risk for damage caused by any of the 'all risks' without the need to insure. Contractors should note, however, that if the employer opts to accept the risk (rather than the sole risk), the employer has the right to refuse payment to the contractor for rectification work to the extent that the loss or damage is due to the negligence of the contractor. If the employer opts for taking the risk only, the contractor should consider whether he should himself insure against possible liability.

Joint Fire Code

Clauses 22FC and 6.3FC of JCT 98 and IFC 98 apply where the Appendix states that the Joint Fire Code applies. The Joint Fire Code means the Joint Code of Practice on the Protection from Fire of Construction Sites and Buildings Undergoing Renovation (published by The Construction Confederation and the Loss Prevention Council). This should be the case if the insurer of the Works requires the employer and the contractor to comply with the Code. The extra clauses place an obligation on both parties to comply with the Code and each indemnifies the other for the consequences of any breach of the Code. There are also time stipulations within which the contractor must carry out remedial measures to rectify a breach. Where the insurer has stated that the Works are a 'large Project' the Code specifies additional requirements.

2.5 Insurance – liquidated damages

Clauses 22D (JCT 98) and 6.3D (IFC 98) have been added to allow the architect to instruct the contractor to obtain a quotation for insurance against delay caused by insurance risks. The amount to be insured is the amount of liquidated damages which the employer would otherwise receive and the clause can only be operated if the Appendix has been completed to that effect and the period of time for which it is to be effective has been inserted. Thus, if the period of time entered is 10 weeks and the contractor is delayed by insurance risks for which he is entitled to an extension for 12 weeks, the employer, if he has operated the clause, can claim an amount equal to liquidated damages for 10 of those weeks from the insurers. If the architect so instructs, the contractor is responsible for accepting the quotation and depositing the policy with the employer. The premium amounts are to be added to the contract sum.

2.6 Sub-contractors

Protection for nominated sub-contractors is provided under JCT 98 clause 22.3 and IFC 98 clause 6.3.3 (named persons as sub-contractors) whereby the contractor or employer, as appropriate, will ensure that the joint names policy either:

- provides for recognition of each nominated sub-contractor as an insured; or
- includes a waiver by the insurers of the right of subrogation against any nominated sub-contractor in respect of specified perils.

The latter is somewhat less than the all risks insurance applicable to the contractor or employer. Domestic sub-contractors are similarly protected except for damage occurring to existing buildings under JCT 98 clause 22C.1 or IFC 98 clause 6.3C.1.

2.7 MW 98 insurance

The insurance position under MW 98 is broadly similar to JCT 98 and IFC 98, but it is much more briefly set down and there are differences, mainly omissions.

The contractor has the same indemnity and insurance liabilities in respect of injury or damage to persons or property, but there is no provision for insurance to cover damage to adjacent property if no one is negligent. Presumably this is because the contract is primarily intended for small Works. The size of the job, however, is no indication of the possibility of such damage and each project should be assessed on its merits and a suitable clause inserted if necessary.

Essentially, clause 6.2 makes the contractor liable for damage to property, except the Works which are the subject of the building contract, to the extent that the damage is due to the contractor's or any sub-contractor's negligence or default. The contractor will be liable to a partial extent if he is partially at fault and damage is caused to surrounding buildings, passing vehicles or an existing building to which the Works are being carried out. The Works are expressly excepted, but if damage is caused to the Works themselves through the contractor's fault, the contractor will be liable, because his obligation is to carry out and complete the Works in accordance with the contract documents. The contractor is obliged to carry appropriate insurance.

If the Works are alterations or extensions to an existing structure, the employer is obliged to insure the Works and the existing structure under the provisions of clause 6.3B. Effectively, the employer and contractor agree that if there is damage caused other than by the contractor's negligence, this clause would deal with the situation [9].

There are only two possibilities in respect of damage to the Works: insurance of new Works by the contractor or insurance of existing structures by the employer. The risks to be insured are identical to specified perils in JCT 98 and the parties should consider whether, in any particular case, more extensive insurance is required. There is no provision for determination after loss or damage if just and equitable and the parties must make their own arrangements in such circumstances. Otherwise the position after damage is the same as under the other contracts. The party responsible for insuring is required to produce evidence of insurance to the other on request but, if the other defaults, there is no provision for either the employer or the contractor to insure, nor can the cost of such premiums be recovered. Since the situation is neither provided for in the determination clauses nor, it is thought, sufficient to allow repudiation at common law, it might be best for the party not in default to take out the appropriate insurance himself, after due notice, and then take the matter to arbitration to claim damages for breach of contract.

It is essential for the architect to read the contract provisions carefully and ensure that the appropriate insurances are in force before the contractor takes possession of the site. It is the architect's duty either to check the contractor's insurance, to ask an expert to do so or to make sure that the employer obtains expert advice [10]. The JCT have produced a detailed guide to the insurance clauses (Practice Note 22); architects, contractors and sub-contractors are advised to obtain copies and give them careful study.

References

1. City of Manchester v Fram Gerrard Ltd (1974) 6 BLR 70.
2. A.M.F. International Ltd v Magnet Bowling Ltd [1968] 2 All ER 789.
3. Gold v Patman & Fotheringham Ltd [1958] 2 All ER 497.
4. Kruger Tissue Industrial Ltd v Frank Galliers Ltd and DMC Industrial Roofing & Cladding Services and H & H Construction (1998) 51 Con LR 1.
5. Computer & Systems Engineering plc v John Lelliot (Ilford) Ltd (1989) The Times 23 May 1989.
6. London Borough of Barking and Dagenham v Stamford Asphalt Co Ltd (1997) 82 BLR 25.
7. Scottish Special Housing Association v Wimpey Construction UK Ltd [1986] 3 All ER 957.
8. English Industrial Estates Corporation v George Wimpey & Co Ltd (1972) 7 BLR 122.
9. National Trust for Places of Historic Interest or Natural Beauty v Haden Young (1994) 72 BLR 1.
10. Pozzolanic Lytag Ltd v Bryan Hobson Associates (1998) CILL 1450.

3 Third parties

3.1 Assignment and sub-letting

Under JCT 98, IFC 98 and MW 98 the contractor has no automatic right to sub-let work. The position is covered by JCT 98 clause 19, and IFC 98 and MW 98 clauses 3.1 and 3.2 (for the latter two, the same numeration in both contracts).

Although assignment and sub-letting are coupled in all three forms of contract, they are, in fact, totally different concepts. Assignment occurs when a right or duty is legally transferred from one party to another. After the transfer, the original party retains no interest in the right or duty transferred. When this is carried out properly, it is termed novation and involves the formation of a new contract.

The parties to a building contract are, of course, the employer and the contractor. In broad terms their rights are to receive a completed building and to receive payment respectively. Their duties echo the rights. The employer's duty is to pay and the contractor's duty is to complete the work. Under the general law a party is entitled to assign the benefit of his rights to a third party, but neither employer nor contractor may assign their duties. An express term in the contract may modify this position. Thus, in the absence of any express term, a contractor may assign his right to payment to a third party in return for money 'up front' in order to enable him to carry out the contract. An employer may assign his right to the completed building to another for payment.

All the three contracts under consideration contain an express term forbidding assignment of rights or duties by either party unless both parties agree. There is no stipulation that a refusal of consent must be reasonable. There appears to be no good reason why the contracts should not be amended to permit assignment of rights and, if a contractor is anxious to raise working capital in this way, it should be brought to the attention of the employer at tender stage. If a contractor or employer attempts to assign rights despite the clause to the contrary, the assignment will be ineffective [1]. The clause will also prevent assignment of the right to damages for breach of contract. However, if the purported assignee cannot sue the contractor, because the assignment is prohibited, the employer may be able to recover damages (presumably on behalf of the purchaser) if it could be seen by the contractor that a subsequent purchaser of the building might suffer loss. This is a difficult point in practice [2].

Clause 19.1.2 applies only when so stated in the Appendix. It allows the employer to assign the right to bring proceedings in his name against the contractor to enforce any term of the contract made for the benefit of the employer. The power may be used if the employer sells or leases the whole of the premises comprising the Works after practical completion. Without this important clause, a future purchaser or lessee of the property has very little chance of obtaining satisfaction through tortuous remedies [3]. It is becoming very common now for the contractor, sub-contractors, suppliers and all the consultants to enter into warranties which themselves contain assignment rights and often very much more onerous conditions than the standard form of contract. The third party is not entitled to dispute enforceable agreements made between the employer and the contractor before the date of the assignment.

Sub-letting

Sub-letting is quite different from assignment. It is the delegation rather than the transfer of a duty. For example, if a contractor sub-lets plumbing work to a sub-contractor, the main contractor remains responsible if the sub-contractor fails to perform or performs badly. If, however, the contractor were able by novation to assign his duty to carry out the contract to a second contractor, it would be the second contractor who would be liable if the work were done badly.

Perhaps in an attempt to clarify the position, clause 19.2.2 provides that even though part of the Works may be sub-let, the contractor remains wholly responsible for carrying out and completing the Works in all respects. A similar provision is included in clause 19.5 in respect of nominated sub-contractors.

All three forms of contract forbid sub-letting of the whole or any part of the Works without the written consent of the architect. There is a proviso that the architect's consent must not be withheld unreasonably. Note that the contractor is only obliged to obtain consent to the fact of sub-letting, not to the sub-contractors themselves. In practice, of course, the architect may well consider that it is reasonable for him to refuse consent unless he knows the names of the sub-contractors. All the contracts now require the contractor to ensure that sub-contracts include a term to entitle the sub-contractor to simple interest at five per cent above Bank of England Base Rate if the contractor fails to pay money due by the final date for payment.

MW 98 has nothing more to say about either assignment or sub-contracting and there is no standard form of sub-contract. JCT 98 and IFC

98 contain further provisions. If the contractor's employment is determined, for any reason, under the main contract, the employment of the sub-contractor under the sub-contract is to be similarly determined. This provision, of course, will only be effective if it is included in the sub-contract because the sub-contractor is not a party to the main contract and cannot be bound by its terms. Contractors should, therefore, take care to use the forms of sub-contract devised for use with the standard main contract forms or ensure that all matters in the main contract which might affect the sub-contract are incorporated in the sub-contract. DOM/1 is the domestic form of sub-contract for use with JCT 98 and IN/SC is the equivalent for use with IFC 98.

Ownership

An attempt is made in JCT 98 clause 19.4.2 and IFC 98 clause 3.2.2 to protect the employer against retention of title clauses preventing ownership of materials passing to the employer. The dangers were highlighted in a case concerning a sub-contractor [4]. This question is discussed in section 1.3. The clauses stipulate that the contractor must include conditions with regard to unfixed materials provided by the sub-contractor. The conditions are set out in the clauses. In essence they are:

- Unfixed goods must not be removed from the Works except with the architect's consent.
- If the value of the goods has been included in a certificate which the employer has paid to the contractor, the goods become the property of the employer and the sub-contractor will not deny it.
- The goods will be the property of the contractor if he has paid the sub-contractor before the employer has paid on a certificate.

If for any reason the contractor does not include such conditions, he is in breach of contract. That is likely to be small comfort to the employer because problems with retention of title are only likely to occur if the main contractor becomes insolvent. In such a case, suing for breach of contract would seem to be a fruitless exercise. The provision may, however, lead to architects asking to inspect sub-contracts to ensure that the conditions have been incorporated. Even if such provisions are included in sub-contracts, they will not be effective against sub-sub-contractors or suppliers to the sub-contractors who may have a retention of title clause in their sub-sub-contracts or contracts of sale. The only way for the employer to be sure is if the provisions are stepped down to the earliest point of supply or if an amendment is made so that the architect is not required to certify for unfixed materials on site. Although one form of contract contains provisions for putting that option into effect, it is unlikely to be popular with contractors and rightly so [5].

IFC 98 goes on to deal with 'named persons as sub-contractors', a provision that will be discussed at length in section 3.2, where nominated sub-contractors under JCT 98 clause 35 will also be considered.

Domestic sub-contractors

JCT 98 contains further important provisions with regard to what it terms 'domestic sub-contractors', in other words sub-contractors who are not nominated. Clause 19.3 provides the employer with a useful alternative to nomination in appropriate cases. The system is that the work to be done must be measured or described adequately in some other way in the bills of quantities so that it can be priced by the contractor. A list of persons or firms is provided in the bills from which the contractor must choose to carry out the work. The contractor has sole discretion in selecting the firm.

The list must contain at least three names. Presumably the architect will have contacted all the firms on the list to make sure that they are willing and able to carry out the work. Either the employer (or the architect acting on his behalf) or the contractor may add more names to the list at any time before a binding sub-contract is entered into in respect of the particular work. The only proviso is that the other party must consent to the additional names. Consent, however, is not to be unreasonably withheld. This provision can be useful to a contractor who wishes to use a firm not included on the original list. It seems that the only reasonable ground for refusing consent, so far as the employer is concerned, would be that the suggested additional firm is not capable of carrying out the work to the standard required by the contract.

Note that the additional names may be added by either party even after the main contract is let. This gives maximum flexibility to the contractor to take advantage of competitive prices.

If at any time before the contractor has entered into a binding sub-contract the number of firms able and willing to carry out the work falls below three:

- The employer and the contractor may agree to add firms so that the list comprises at least three.
- Alternatively the contractor may carry out the work himself, and in so doing, he may sub-let the work to any sub-contractor of his choice provided the architect gives his consent.

A sub-contractor chosen from the list by the contractor becomes a domestic sub-contractor. The employer will not, thereafter, be concerned with any problems of delay, financial claims or determination of employment. These are solely the concern of the contractor and the sub-contractor involved. The contractor, of course, may be able to found a claim on events relating to the sub-contract work as if the work was carried out by the contractor's own operatives (for example, extension of time for exceptionally adverse weather conditions). The contractor, however, remains responsible for the work of his domestic sub-contractors, and for any defects therein, to the employer.

Sub-contract

This is not the place to discuss the forms of sub-contract, but there are some general points which deserve mention if only to emphasise the importance of having a sub-contract which sets out the rights and duties of the parties in a realistic manner.

Many problems between main and sub-contractors arise because a project is not ready for a sub-contractor to commence work on the date anticipated, or the sub-contractor is prevented from maintaining reasonably expected progress due to delays for other reasons. If the commencement of the sub-contract is delayed, the sub-contractor may say that he wants more money or even that he cannot, at a later date, fit the work into his programme. Whether the sub-contractor can lawfully take that stance depends upon the terms of the sub-contract. If there are no express terms or agreement about the date for commencement, the law will not imply that the contractor must have the site ready for the sub-contractor by any particular date. The most which will be implied is that the contractor must have the site ready within a reasonable time [6].

Similarly, in the absence of an express term of the sub-contract, there will be no implied term that the contractor must ensure that there is sufficient work available to enable the sub-contractor to maintain reasonable progress and execute the sub-contract Works in an efficient and economic manner [7]. However, it is now established that sub-contract form DOM/1 simply requires the sub-contractor to finish his work by the date fixed in the sub-contract. Provided he does that, he may plan and perform the work as he pleases [8]. Thus, if the sub-contract contains a term requiring the sub-contractor to proceed with the sub-contract Works at such times and in such manner as the contractor directs, the sub-contractor will be required to do exactly as stated. In many instances, the contractor will supply a sub-contractor with a programme, but it will seldom be a contract document. Only if it is a contract document will the sub-contractor be entitled to work to it and an instruction to work differently would constitute a variation [9].

There is comfort for sub-contractors in recent case law developments. In the absence of a warranty, employers will find it very difficult to take direct action against sub-contractors in tort. The only exceptions are likely to be if the sub-contractor's negligence causes damage to property other than the Works or if there is a danger or imminent danger to health or safety. Where there is a warranty, its terms will not necessarily constitute the whole of the sub-contractor's liability and the courts are likely to find a concurrent liability in tort [10]. This liability may sometimes exceed the sub-contractor's liability in contract [11].

3.2 Nominated sub-contractors and suppliers

An employer who nominates a sub-contractor is often accused of trying to have everything his own way. He wants to stipulate who the contractor must employ to carry out certain work, but at the same time, the employer wants no responsibility. In practice, nomination is not quite like that; in fact, it often causes more problems for the employer than for the contractor. Nomination is best avoided if possible. Sometimes it is essential that a specialist sub-contractor be nominated for particular work. On those occasions, all parties should read the fine print.

MW 98 makes no provision for nomination. There are ways of overcoming the difficulty which will be discussed later, but this form of contract is not intended for use if the employer wishes to nominate. IFC 98 makes no provision for nomination, but instead, for something called naming. This is not just a matter of terminology. The provision is vastly different from the nomination procedures in JCT 98. JCT 98 deals with the nomination of sub-contractors in clause 35 and with suppliers in clause 36. Clause 35 is long and, whatever the armchair pundits may say, complex in wording and operation. What follows is no substitute for the tedious job of reading the clause and digesting its many provisions.

Nomination can take place in any of eight ways (JCT 98 clause 35.1). The architect may either use a prime cost sum or name the sub-contractor:

- in the bills of quantities;
- in an instruction regarding the expenditure of a provisional sum;
- in an instruction requiring a variation under clause 13.2;
- by agreement with the contractor.

If the architect agrees and the type of work is that which the contractor normally carries out, the contractor may be allowed to submit a tender.

Nomination procedure

The appropriate forms are:

NSC/T Tender;
NSC/W Warranty;
NSC/N Nomination; and
NSC/A Articles, which incorporates NSC/C Conditions.

NSC/T is divided into Parts 1, 2 and 3 and Part 3 is Particular Conditions to be agreed. The procedure is basically as follows:

- The architect sends the completed NSC/T Part 1 to the prospective sub-contractor together with a blank Part 2, the drawings and/or specification and/or bills of quantities (the 'numbered tender documents') and the Appendix to the main contract as it is envisaged to be completed. In addition, the architect must include a copy of NSC/W with the contract details completed.
- The sub-contractor must complete NSC/W and Part 2 and return them to the architect.
- The employer signs Part 2 as approved and enters into NSC/W.
- The architect then sends to the contractor a nomination instruction NSC/N, NSC/T Parts 1 and 2, the numbered tender documents and NSC/W. At the same time, the architect must send to the sub-contractor a copy of NSC/N, the completed NSC/W and the main contract Appendix as completed.
- The contractor and sub-contractor are to agree the Particular Conditions (Part 3) and to enter into a sub-contract on NSC/A (which incorporates conditions NSC/C) within 10 working days of receipt of the nomination instruction and to send the architect a copy of the completed NSC/A.

If the contractor fails to enter into NSC/A, he must inform the architect either:

- *the date he expects to complete NSC/A* and the architect may fix whatever new date for completion he considers reasonable; or
- *that the failure is due to other matters which must be specified* and the architect may:
  - issue further instructions to allow completion;
  - cancel the nomination and omit the work;
  - cancel the nomination and nominate another sub-contractor;
  - if he does not consider the matters justify the failure to complete NSC/A, notify the contractor accordingly who must then agree the Particular Conditions and enter into the sub-contract.

There appear to be some difficulties in operating the last option, because the contract cannot impose an obligation on two parties to agree. In any event, the sub-contractor is expressly entitled to withdraw his tender under NSC/T Part 2, either within seven days of notification of the identity of the main contractor or if he is 'for good reasons' unable to agree Part 3 Particular Conditions. The general law would entitle him to withdraw his tender at any time before acceptance whatever he may have said about leaving it open for any particular period unless payment or some other consideration was given by the employer.

The contractor may make reasonable objection within seven days of receipt of the architect's nomination instruction.

Payment

Amounts due to nominated sub-contractors are to be included in interim certificates and the architect must direct the contractor regarding the amounts and the sub-contractors to whom payment is to be made. The contractor must provide the architect with reasonable proof of payment before the issue of the next certificate. If the contractor fails to provide the proof, the employer may, but if NSC/W is in operation must pay the sub-contractor direct and deduct the amount from the sum due to the contractor on the next certificate. The employer's power to pay direct is subject to the architect certifying that proof has not been provided. The direct payment provision ceases to have effect if the contractor becomes insolvent.

Extension of time

The sub-contractor is entitled to an extension of time on much the same grounds as the contractor under the main contract together with the contractor's acts, omissions or defaults. The contractor, however, must obtain the architect's consent before granting an extension. If the sub-contractor fails to complete his work during the sub-contract or extended period, the contractor must notify the architect who, if he is satisfied that the extension of time procedures have been properly applied, must issue a certificate of non-completion. The issue of this certificate enables the contractor to recover damages from the sub-contractor which will, in appropriate instances, include but not be limited to liquidated damages which the contractor is liable to pay to the employer under the main contract.

Defects

The architect must certify when practical completion of the sub-contract Works has taken place (clause 35.16). At any time thereafter and no later than 12 months later, the architect issues an interim certificate including the final payment to the nominated sub-contractor. There is a proviso that the sub-contractor must have furnished all documents necessary to calculate the final sum and he must have remedied any defects in accordance with the sub-contract terms. Once the contractor has discharged this final payment, the position is broadly that any other defects in the sub-contract work before the final certificate may be the contractor's responsibility. If the sub-contractor fails to remedy them, the employer may nominate someone else and, if the employer cannot recover the cost from the original sub-contractor, it can be recovered from the main contractor. This places a heavy burden of sub-contract supervision on the contractor's shoulders.

The contractor is generally responsible to the employer for any defect in the sub-contract Works, but clause 35.21 expressly excludes responsibility for any element of design or selection of materials which has been left to the nominated sub-contractor. Moreover, if the sub-contractor manages to limit his liability under NSC/W, the contractor's liability is likewise limited.

Renomination

Problems arise if the sub-contractor stops work before completion. The provisions are rather complicated, but basically the architect is under a duty to nominate another firm to complete the work and, if necessary, deal with defects in the work already carried out [12]. If the new nomination cannot complete the work within the original sub-contract period, the contractor is entitled to an extension of time. If the original sub-contractor withdraws, the contractor has neither the right nor the duty to do the work himself [13].

Renomination will become necessary if:

- The contractor is of the opinion that the nominated sub-contractor has defaulted in certain specified matters.
- The sub-contractor becomes insolvent.
- The sub-contractor's employment has been determined for corruption.
- The sub-contractor determines his own employment due to certain specified defaults of the contractor.

If the contractor notifies the architect that the sub-contractor has defaulted in any of the respects listed in the sub-contract NSC/C clause 29, he must include the sub-contractor's observations. If the architect is reasonably of the opinion that the sub-contractor has defaulted, he must instruct the contractor to serve notice on the sub-contractor specifying the default. The architect may state that the contractor must obtain a further instruction before determining the sub-contractor's employment. The contractor must notify the architect when determination has taken place. The architect must renominate within a reasonable time. He is entitled to invite tenders for the work in order to achieve a fair price. The renomination must include the doing or redoing of work and the supply or resupply of materials. If the grounds for determination are failure to comply with the contractor's written notice to remove defective work or materials or failure to correct defects, the contractor is entitled to agree the price of the substituted sub-contractor. The contractor, however, cannot unreasonably withhold his agreement. This provision is important because if the contractor is unable to recover from the original sub-contractor the extra cost of the substituted sub-contractor, the contractor must pay the difference.

In the case of what might be collectively termed insolvency or in the case of corruption, determination is automatic or as a requirement of the employer respectively. Legal advice must always be sought in the case of determination due to insolvency because the position is very complicated. If the receiver, administrative receiver or administrator is willing to carry on the sub-contract, the architect may postpone the renomination if it will not prejudice the interests of either the employer, the contractor or any other sub-contractor.

If the contractor causes the sub-contractor to determine by reason of the contractor's default as specified in the sub-contract clause 30, the architect must renominate. When the amount due to the substituted sub-contractor has been certified, the employer may deduct the extra he has had to pay from any money due to the contractor or he may recover it as a debt.

Suppliers

The provisions relating to nomination of suppliers are relatively short. The principal points to note are that the architect must only nominate a supplier who is prepared to enter into a contract of sale on the terms set out in clause 36.4 unless the architect and the contractor agree otherwise. There are grounds for varying the delivery programme. This overcomes the technical point that if a nominated supplier introduces such grounds into his own contract of sale, they would be contrary to clause 36.4.9 which provides that no terms in such a contract may override, modify or affect the terms set out in clause 36.4.

Contractors should note that they are not obliged to enter into a contract of sale which restricts the supplier's liability to the contractor unless the architect has given written approval. If he gives such approval, the contractor's liability to the employer is similarly restricted.

Named persons

IFC 98 provides, in clause 3.3, for persons to be 'named'. There are two situations:

- where work is included in the contract documents and priced by the contractor to be carried out by a person named in the documents;
- where there is a provisional sum and the architect issues an instruction naming a person to carry out the work it represents.

In the first situation, the contractor must enter into a sub-contract (NAM/SC) with the named person within 21 days of entering into the main contract. The main contract will become binding when the employer accepts the contractor's tender. There is, therefore, very little time. A substantial quantity of documentation is involved which is obviously intended to be part of the tender documentation and, thereafter, part of the contract documents. That presupposes that all sub-contract tenders will be received well before the main contract tenders are invited. This first option is not something which will appeal to an architect when time is pressing. He will be more likely to take the second option.

If the contractor cannot enter into a sub-contract in accordance with the particulars in the contract documents, he must notify the architect. If the architect is satisfied that the particulars have prevented the execution of the sub-contract, he has three options:

- alter the particulars, creating a variation;
- omit the work, creating a variation;
- omit the work and substitute a provisional sum for which an instruction is required.

Note that the architect can only alter the particulars in so far as they are not contract terms. Only the parties to a contract can agree to vary its terms. If the work is omitted in its entirety, the employer may use his own directly employed labour to carry it out under clause 3.11.

In the second situation, the instruction on a provisional sum can arise:

- after the contractor has notified the architect that the particulars are preventing execution of the sub-contract;
- if the provisional sum is substituted before the sub-contract is signed;
- if there is already a provisional sum in the contract documents.

The instruction must name a person, describe the work and include all the complex documentation (NAM/T). In this instance, the contractor has 14 days from the date of the instruction in which to make a reasonable objection. If a reasonable objection is made, the contract leaves the position open. Presumably the architect must name another firm and so on. The procedure could be protracted and, again, the architect has no power to grant an extension of time (which may thus become 'at large'; see section 6.1). If there is no objection, the contractor must enter into a sub-contract with the named person.

There is a form of employer/sub-contractor agreement, similar in intent to NSC/W, which should be completed (ESA/1). This provides the only redress for the employer if the sub-contractor fails in the design or selection of materials for which he has undertaken responsibility. The contractor has no liability in these areas. What is worse, from the employer's point of view, no other sub-contractor is to be so liable since the employer's only redress in the case of ordinary (clause 3.2) sub-contractors is through the main contractor. In general, however, the contractor is responsible for the normal workmanship and materials aspects of a named person's work.

Determination

If the named person's employment is determined, the architect may:

- name another person;
- instruct the contractor to make his own arrangements to do the work;
- omit the work still to be completed.

In the last case, the employer may employ his own labour to carry out the work.

The consequences of determination are complex, depending upon whether the work was originally included in the contract documents or arises as a result of an instruction regarding the expenditure of a provisional sum. Briefly, in the first case if the architect has exercised his option of naming another person, the contractor is entitled to an extension of time, but not a claim for loss and/or expense. In all other circumstances, the contractor is entitled to both time and money. It is important to note, however, that, if the determination occurs through the contractor's fault, he can have none of these benefits.

The contractor must take whatever action is necessary within reason to recover from the named person any loss sustained by the employer as a result of the determination. The contractor need not take any legal proceedings unless the employer agrees to pay any legal costs incurred.

MW 98

It is possible effectively to nominate sub-contractors in MW 98 by:

- naming the firm in the contract documents;
- naming the firm in an instruction regarding a provisional sum (clause 3.7);
- including a specially worded nomination clause in the contract.

The big problem with the first two options is that there is no provision for the consequences, which case law has shown can be considerable. To include a special clause is probably the most satisfactory, but it is counter to the spirit of this simple form of contract and is probably sufficient to class the whole contract as the employer's written standard terms of business for the purpose of the Unfair Contract Terms Act 1977.

Points

A final point in regard to nominated sub-contractors. In general terms it can be said that the inclusion of a 'pay when paid' clause in the sub-contract will be effective against the sub-contractor, who has completed his work, if the employer becomes insolvent before paying to the main contractor sums due to the sub-contractors [14]. However, under the provisions of NSC/C, which has no pay-when-paid clause, if the main and sub-contracts are determined before the architect has certified under the main contract, the sub-contractor can still recover monies to which he is clearly entitled [15]. It should be noted that the Housing Grants, Construction and Regeneration Act 1996 section 113 states that payment provisions which attempt to make payment dependent on receipt of payment from a third party (pay-when-paid provisions) are ineffective except in the case of third party insolvency.

3.3 Employer's licensees

When the contractor is in possession of the site he is said to have a licence to be there in order to carry out his contractual obligations. Ideally, those obligations should embrace every aspect of the Works and the construction should be entirely under his direction and control. The drawings and bills of quantities should be complete and accurate and the architect should only be required to make occasional visits to keep an eye on progress. Life, certainly a contractor's life, is rarely like that.

Quite apart from the contractor's own shortcomings, inaccuracies in drawings and a steady stream of variations, JCT 98 and IFC 98 provide for the possibility of other contractors, quite separately engaged by the employer, arriving on site and carrying out part of the Works. The clauses governing the situation are 29 and 3.11 respectively and the separately engaged contractors are termed 'employer's licensees'.

The clauses are similar in intention although not quite identical in wording. Two situations are envisaged:

- where the contract documents provide for work not forming part of the contract to be carried out by the employer or by persons employed or engaged by him;
- where the contract documents do not make any provision for work not forming part of the contract.

The only sensible interpretation to give to 'work not forming part of the contract' is that it refers to work which the contractor has no obligation to carry out nor right to receive payment for. In the first situation, the detail in the contract documents must be sufficient to enable the contractor to inform himself precisely of what is to be done and when it is to be carried out. Items of attendance must be specified so that the contractor has the opportunity, at tender stage, to assess the cost to him. Provided this is done, the contractor has no option, as the work proceeds, but to allow the directly employed contractors to do their work.

In the second situation, there is no mention of the work in the contract documents and if the employer wishes to employ others, he must first seek the contractor's consent. The contract stipulates that the contractor must not unreasonably withhold his consent. What is reasonable will depend upon circumstances. The contractor will require to be informed about the extent of the work, its timing and probably the identity of the person or firm to carry it out.

Reasonable grounds for refusal of consent could be that the contractor has suffered bad experiences working alongside such person or firm in the past or that to allow the work would seriously interfere with progress. The wise contractor will carefully consider all aspects and obtain written assurances from the employer (not the architect) before giving consent.

Clearly any work done by others during the time the main contract is in progress will cause some disturbance, but there are remedies available which will be discussed later. If work is fully detailed in the contract documents, but the employer wishes to change the work in some way, it immediately falls into the second category. There is no provision for varying this work. The variation clauses in the contract refer only to contract work, i.e. that to be executed by the contractor. In these circumstances the employer must seek the contractor's consent just as though none of the work was mentioned in the documents.

The final part of the clause makes clear that directly employed contractors are not to be considered as sub-contractors but as persons for whom the employer is deemed to be responsible for the purposes of the insurances clauses. This means that the employer must make certain that they are properly insured and, if damage occurs through their negligence, the employer will be liable although he may be able to pass such liability on to his directly engaged contractors.

The employer and the contractor must take care that all these employer's licensees have a separate contract with the employer and they must be paid directly by the employer and not, under any circumstances, through the contractor. If the contractor allows himself to become the channel for payment and instructions to such persons, he runs the risk that they may be considered sub-contractors for whom he is responsible.

Disadvantages

In practice, there are few advantages and very many disadvantages for the employer in bringing persons, other than the contractor, onto the site. Sometimes the employer may wish to do so because, for example, they are his own employees in the case of a local authority or he may have a special relationship with them in the case of sculptors, graphic designers or landscape contractors, or it may simply be that he wants to have complete control over them which would not be the case if he nominated or named sub-contractors.

Once on the site, or even before entering the site, the employer's licensees will inevitably cause delay and disruption. In the case of a small item such as a specially designed memorial plaque to be fixed complete with unveiling apparatus in the entrance hall of a public building, the disturbance may be so slight as to be insignificant. The work not forming part of the contract, however, may be quite substantial and scheduled to be carried out quite early in the job. Take, for example, the case of special equipment which may necessitate the early supply of baseplates, conduits, etc. which the contractor may have to build in or which the specialist firm insists on fixing itself. If it is late or even if it is on time, severe disruption may result.

Fortunately, the contract provides remedies for the contractor. This of course is the disadvantage as far as the employer is concerned. The extension of time clause (JCT 98 clause 25 and IFC 98 clause 2.4) makes the execution of the work, or failure to execute work not forming part of the contract, a ground for extension of the contract period. Note that it is the execution of the work, quite correctly and at the right time, as well as failure to execute which is the ground. The implication is clear: even if the work is carefully detailed in the contract documents and the employer's licensees carry it out without any fault, a ground for extension exists. Of course, the architect is entitled to take all the circumstances into account, but it seems that he cannot avoid granting some extension.

Where the work is not previously detailed and only carried out with the contractor's consent, the contractor will certainly be entitled to an extension to cover the carrying out of the work if thereby the completion date is exceeded. Moreover, the provisions for loss and/or expense (JCT 98 clause 26 and IFC 98 clause 4.12) repeat the wording precisely as ground for a financial claim. Since the presence of other contractors on the site will almost invariably cause disruption to regular progress to some extent, the inclusion of employer's licensees may amount to much the same as a blank cheque from the employer.

The contractor's remedies, and employer's problems, do not end there. If the execution or failure to execute such work causes a delay for a period to be named in the Appendix, in the case of JCT 98 clause 28.2, or of one month, in the case of IFC 98 clause 7.5.3, the contractor may determine his employment under the contract with all that entails for the employer (see section 7.2).

Another aspect, which is closely related to employer licensees, is the situation if the employer undertakes to supply any materials or goods for the Works. There is no specific provision in the contract for the employer to undertake direct supply, therefore an additional clause should be inserted.

It should be noted, however, that the extension of time, loss and/or expense and determination clauses all contain reference to supply by the employer in much the same terms as to directly employed contractors. So if, for example, the employer states in the contract that he will supply all the ironmongery for the project, perhaps because he has a cheap source of supply, he must supply ironmongery of the correct type, in the correct quantities, of the correct quality, at the correct time and without any defects, if he is to avoid any claim from the contractor [16].

The contractor's letter giving consent under clause 29.2 (JCT 98) or clause 3.11 (IFC 98) is most important and it is suggested that it should be worded somewhat along the following lines:

> *To Employer*
>
> *Dear Sir,*
>
> *We are in receipt of your letter referring to the engagement of* (insert name) *in accordance with clause 29.2* (substitute 3.11 when using IFC 98) *of the conditions of contract. The work is not detailed in the contracts bills (or specification), but we understand that it will consist of* (details) *and it will commence on site on* (date) *reaching completion on* (date).
>
> *We give our consent to the work subject to your written confirmation that you recognise our entitlement to extension of time and loss and/or expense in consequence.*
>
> *Yours faithfully,*

It is clear that an employer making use of the provisions to employ other contractors on the site places himself in a very vulnerable position. He would be well advised to allow all the work to be carried out through the contractor or, at least, arrange for any additional work to be done after the contractor has reached practical completion stage and left the site. Most contractors would also be well pleased with that arrangement despite generous remedies available in the contract.

MW 98

MW 98 makes no mention of employer's licensees. Any such arrangement must be with the consent of the contractor and, of course, in this case, he need not be reasonable in withholding his consent. If he does consent, extension of time would be granted under clause 2.2, but there is no contractual provision for loss and/or expense. It may be difficult to sue at common law because of the difficulty of proving breach of contract since the contractor, by his consent, has agreed. The answer would seem to be for the contractor to consent on condition that the employer makes special provision to reimburse any loss and/or expense. Powers of determination, however, are available under clause 7.3.

3.4 Statutory provisions

Every person or firm has a duty to comply with requirements laid down by statute, that is by Act of Parliament. It seems clear that a contractor's duty to comply with statutory requirements will prevail over any express contractual obligation [17].

Clauses dealing with statutory obligations are found in JCT 98 clause 6, IFC 98 clause 5 and MW 98 clause 5. JCT 98 and IFC 98 provisions are very similar and they will be considered first.

The meaning of 'statutory requirements' is helpfully defined as being requirements of 'any Act of Parliament, any instrument, rule or order made under any Act of Parliament, or any regulation or by-law of any local authority or any statutory undertaker' relative to the Works (JCT 98 clause 6.1.1). IFC 98 clause 5.1 has a somewhat shorter definition which amounts to the same thing. Statutory instruments are usually made by a Secretary of State and the most important, from the contractor's point of view, will probably be the Building Regulations 1993 and the Construction (Design and Management) Regulations 1994, usually referred to as the CDM Regulations, which are considered below.

The contractor must comply with all statutory requirements and give all notices which may be required by them. He is also responsible for paying all fees and charges legally demandable in respect of the Works. He is entitled to have such amounts added to the contract sum unless they are already provided for in the contract or unless, in the case of JCT 98, they arise in respect of work or materials carried out or supplied by a local authority or statutory undertaker in a nominated capacity.

Indemnity

JCT 98, but not IFC 98, contains an indemnity provision which deserves attention. Not only must the contractor pay charges as above, but he agrees to indemnify the employer against liability in respect of such charges. Therefore, if the contractor fails to pay as legally required, he assumes liability on behalf of the employer. Such liability might well extend to undoing work already done, delays or fines. This is an onerous provision which might easily be overlooked. Its purpose is to keep the employer safe from damage or loss and its effect would be rather broader than the position under IFC 98, where the employer's remedy for the contractor's failure to pay would be to sue for damages for breach of contract.

Indemnity clauses tend to be interpreted by the courts against the person relying on them. In particular, they must be drafted in very precise terms if they are to include indemnity against the consequences of the employer's own negligence [18]. This may give the contractor some relief, but he should note that the time during which he remains liable under an indemnity clause does not begin to run until the liability of the employer has been established, usually by the court.

Divergence

The contractor has a duty to notify the architect immediately if he finds a divergence between statutory requirements and the contract documents or architect's instructions. The contractor has no obligation to search for such divergences (see section 1.1). When the architect receives the notice, or if the architect finds a divergence himself, JCT 98 clause 6.1.3 gives him just seven days to issue an instruction about the divergence. If the architect fails to meet this deadline, the contractor would appear to have clear grounds for an extension of time under clause 25.4.6.2 and reimbursement of loss and/or expense under clause 26.2.1.2. IFC 98 makes no mention of architect's instructions in this regard, but obviously instructions must be given within a reasonable time or the contractor would have similar claims.

Provided that the contractor notifies the architect if he finds a divergence and otherwise carries out the work in accordance with the contract documents and any other drawings and instructions issued by the architect, both forms of contract give the contractor a valuable safeguard. He is not to be held liable to the employer if the Works do not comply with statutory requirements. The contract is still liable for compliance as far as, for example, the local authority is concerned, but he should be able to recover his costs for rectifying such work from the employer. Thus, if the contractor fails to find a divergence, he is able to escape liability by virtue of this clause.

Emergency

In practice, the contractors should beware of carrying out instructions from anyone other than the architect. If the building control officer directs that work does not comply with the Building Regulations the only safe course is to refer the matter immediately to the architect. The only exception to this is in the case of an emergency. If the matter really is urgent, the contractor may carry out the necessary work immediately provided that he:

- notifies the architect forthwith of the steps he is taking; and
- supplies only sufficient materials and carries out just enough work to ensure compliance.

The contractor is then entitled to payment for what he has done just as though the architect had issued an instruction.

MW 98 contains broadly similar provisions in clause 5.1. The contractor must pay all charges and comply with all statutes, instruments, etc. He must notify divergences which he finds, after which he has no liability for non-compliance as far as the employer is concerned. There is no provision for any indemnity similar to JCT 98, nor is there provision for emergency compliance. The contractor must, therefore, report any urgent matters to the architect. In circumstances where the matter really cannot wait – for example, if there is an element of danger involved – it is thought that the contractor's obligation to comply with statutory requirements must prevail and he would be entitled to claim reimbursement from the employer for emergency work. In practice, he would have difficulty in persuading the architect that he was not in breach for failing to notify an obvious divergence, or negligent in construction. MW 98 provisions are quite brief, but adequate for its sort of project.

CDM Regulations

Breach of the Construction (Design and Management) Regulations 1994 is a criminal offence; however, except for two instances, a breach will not give rise to civil liability. Therefore, one person cannot normally sue another for breach of the Regulations. Compliance with the Regulations is a contractual duty so that breach of the Regulations is also a breach of contract. The important clauses are mainly clause 6A in JCT 98, clause 5.7 in IFC 98 and clauses 5.6 to 5.9 in MW 98. The clauses in all the contracts are very similar and it is only necessary to look at JCT 98 in any detail.

Article 6.1 in JCT 98 assumes that the architect will be the planning supervisor under the Regulations. If one looks carefully, it is possible to see the word 'or' enabling the user to insert an alternative name. It is not at all certain that every architect wants to take on the role. Article 6.2 records the name of the principal contractor. That will almost always be the contractor under the contract.

There are grounds for determination (failure to comply with the Regulations) in the list in both employer and contractor determination clauses (clauses 27 and 28).

Three points are very significant and they should be read together. Clause 6A has been inserted to provide that the employer 'shall ensure' that the planning supervisor carries out all his duties under the Regulations and that, where the principal contractor is not the contractor, he also will carry out his duties in accordance with the Regulations. There are also provisions that the contractor, if he is the principal contractor, will comply with the Regulations. The contractor must also ensure that any subcontractor provides necessary information. Compliance or non-compliance by the employer with clause 6A.1 is a 'relevant event' and a 'matter' under clauses 25 and 26. Lest the significance escapes anyone, what this means is that the employer must ensure that the planning supervisor performs correctly and if he does not, or even if he does, any resultant delay or disruption will give entitlement to extension of time and loss and/or expense. This may well be a most fruitful source of claims for contractors. Every contract administrator's instruction potentially carries a health and safety implication which should be addressed under the Regulations. The

Regulations impose a formidable list of duties on the planning supervisor. Most of them are to be found in Regulations 14 and 15. Some of these duties must be carried out before work is commenced on site. If necessary actions delay the issue of an architect's instruction or once issued delay its execution, the contractor will be able to claim.

Part of clause 42 (performance specified work) makes clear that additional procedures must be operated between contractor and planning supervisor before the contractor provides the architect with his design. To the extent a contractor may be delayed, claims may arise.

There may be occasions when the Regulations do not fully apply to the Works as described in the contract. If the situation changes due to the issue of an architect's instruction or some other cause, the employer may be faced with substantial delay as appointments of planning supervisor and principal contractor are made and appropriate duties are carried out under the full Regulations.

Practice Note 27 has been published. It is very helpful, but it should be read with care and in the knowledge that it has no legal weight. The way in which the clauses will work in practice will become clear in due course. It is certain that the key factor will be for architects, planning supervisors and principal contractors to structure their administrative procedures very carefully if they are to avoid becoming in breach of their contractual obligations.

Statutory undertakers

A statutory undertaker is any body, such as an electricity supplier or gas supplier, which derives its authority from statute. There tends to be some misunderstanding of the position of such a body in relation to the contract. JCT 98 clause 6.3 makes clear that a local authority or statutory undertaker carrying out work as part of its statutory obligations is not a sub-contractor. This appears to be generally understood and, although not expressly mentioned in either IFC 98 or MW 98, the position must be the same under those forms also. The result is that the statutory undertaker has no contractual liability to either employer or contractor when it is using statutory powers. If such a body enters site to lay cables, make connections, etc. within its statutory powers, it may cause severe disruption or delay. All three forms will provide some remedy for the contractor in terms of extension of time, but the contractor cannot contractually recover any loss and/or expense in respect of the disruption either from the statutory undertaker or from the employer.

The position is quite different if the statutory authority is carrying out work on site which is not a statutory obligation. For example, the electricity supplier may be wiring a house as would any electrical sub-contractor. In those circumstances, the authority:

- may have a contract with the contractor and be in the position of a nominated, named or domestic sub-contractor under JCT 98 clause 35 or 19, IFC 98 clause 3.2 or 3.3 or MW 98 clause 3.2;
- may have a contract with the employer and be licensee under JCT 98 clause 29 or IFC 98 clause 3.11.

In the first instance, the contractor's remedy for disruption by the statutory authority would be against the authority itself under the terms of the sub-contract. In addition, depending upon whether the authority was a nominated or named sub-contractor, there would be other important safeguards under the terms of the main contract (see section 3.2). In the latter instance, if the authority were a licensee, the contractor's remedies would be against the employer under the main contract.

It is, therefore, of the utmost importance to the contractor that he correctly identifies the true role of the statutory undertaker. It is generally fairly clear if the undertaker is a sub-contractor, but it may be the subject of dispute whether it is an employer's licensee [19].

A contractor faced with major losses due to the activities of a statutory authority carrying out its statutory obligations may not be totally without remedy if he can show that his loss was caused by the authority's negligence. In those circumstances, the contractor should consult his solicitor to assess the chances of a successful action in tort.

References

1. Helstan Securities Ltd v Hertfordshire County Council (1978) 20 BLR 70.
2. St Martin's Property Corporation Ltd and St Martin's Property Investments Ltd v Sir Robert McAlpine & Sons Ltd, and Linden Gardens Trust Ltd v Linesta Sludge Disposals Ltd (1993) 63 BLR 1; Darlington Borough Council v Wiltshire Northern Ltd (1994) 69 BLR 1.
3. Murphy v Brentwood District Council (1990) 50 BLR 1.
4. Dawber Williamson Roofing Ltd v Humberside County Council (1979) 14 BLR 70.
5. ACA Form of Building Agreement 1984 edition (1998 revision).
6. Piggott Construction Co Ltd v W J Crowe Ltd (1961) 27 DLR (2d) 258.
7. Martin Grant & Co Ltd v Sir Lindsay Parkinson & Co Ltd (1984) 3 Con LR 12.
8. Piggott Foundations Ltd v Shepherd Construction Ltd (1993) 67 BLR 48.
9. Kitson Sheet Metal Ltd & Another v Matthew Hall Mechanical & Electrical Engineering Ltd (1989) 17 Con LR 116.
10. Henderson v Merritt Syndicates (1994) 69 BLR 26.
11. Holt and Another v Payne Skillington and Another The Times 22 December 1995.
12. Rhuddlan Borough Council v Fairclough Building Ltd (1985) 3 Con LR 38.
13. North-West Metropolitan Hospital Board v T A Bickerton & Son Ltd [1970] 1 All ER 1039.
14. A Davies & Co (Shopfitters) Ltd v William Old Ltd (1969) 67 LGR 395.
15. Scobie & McIntosh Ltd v Clayton Bowmore Ltd (1990) CILL 565.
16. Thomas Bates & Sons Ltd v Thurrock Borough Council (1975) CA 22 October.

17. Street v Sibbabridge Ltd (1980) unreported.
18. Walters v Whessoe Ltd and Shell Refining Co Ltd (1960) 6 BLR 24.
19. Henry Boot Construction Ltd v Central Lancashire New Town Development Corporation (1980) 15 BLR 1.

4 Work in progress

4.1 Setting-out

In an effort to start the job and, having started it, maintaining reasonable progress, the contractor sometimes assumes more than his fair share of responsibility. An example of this may be observed in the contractor's attitude to setting-out when faced with wholly or partially inadequate drawings. Very often, he does his best on the information available when he should be seeking to protect his own position.

Setting-out is covered by JCT 98 clause 7 and IFC 98 clause 3.9. MW 98 makes no mention of setting-out, but it is thought that a term is to be implied on similar lines to the express terms in the other contracts. JCT 98 and IFC 98 are almost, but not quite, identical. JCT 98 states that the architect 'shall' (meaning 'must') determine any levels and provide accurately dimensioned drawings with enough information to enable the contractor to set out the Works at ground level. IFC 98 omits the reference to ground level.

Clearly, the contractor may set out the Works below ground level, at ground level and again at higher levels depending on the precise configuration of the building involved. The architect's obligation, however, is to relate the setting-out dimension to ground level when JCT 98 is used.

It is the contractor's responsibility to set out the Works accurately and if he makes any mistake, he must amend it at his own cost. For example, a contractor may set out a building in such a way that the dimensions of the building work properly in relation to all the other building dimensions, but the building as a whole is in the wrong position on the site.

This kind of error may only become obvious when site works are in progress. Clearly the contractor may have carried out many thousands of pounds worth of work by that time. The contractor's obligation seems to be to tear down what he has built and start again, setting it out correctly this time. Certainly, where someone is in breach of contract, the other party is entitled to receive an amount of money which would put him in the same position (so far as money could do it) as if the contract had been correctly carried out [1]. In the case of incorrect setting-out, that would mean the cost of demolition and re-erection. However, this very strict and draconian approach would be modified in practice by the courts or an arbitrator and if the cost of rectification was out of all proportion to the benefit to be gained, rectification would not be supported and a nominal sum might be awarded instead [2]. All the circumstances would have to be taken into account including whether the injured party intended to have the matter put right or simply pocket the money.

If the building encroaches onto neighbouring land, it amounts to trespass and the contractor would be liable for damages if sued directly by the neighbour or as third party if the neighbour sued the employer.

Fortunately, most errors are detected quite early, often while the building is still in the foundation stage and the damage can be limited. What of the situation where the contractor sets out wrongly so that the employer has substantially more building than he expected? There was a case some years ago in which the building, a school, was half a metre longer than shown on the drawings. The error was not discovered until the architect designed a floor tiling pattern and found that it would not fit. On the face of things, the employer had gained rather than lost by the contractor's error. In such an instance it would probably be unreasonable to expect the contractor to correct the error, although that is his strict obligation. The contractor would certainly have no claim for reimbursement for additional work and materials used. Moreover, the employer would probably be able to claim that he would be involved in additional maintenance expense for the life of the building.

It is probably to overcome such problems that both contracts provide that the architect, with the employer's consent, can instruct the contractor not to amend errors in setting-out and make an appropriate deduction from the contract sum. There is no guidance in the contract regarding what might constitute an appropriate deduction, but the situation is similar to that created when the employer opts not to have defects corrected at the end of the defects liability period (see section 7.1) and an appropriate deduction might be the cost to the contractor of rectification of the defective setting-out [3]. It is more likely that a much smaller sum would be indicated. In the case mentioned above, the cost would be substantial and it is doubtful if a court would enforce the point. More likely, the extra cost of maintenance and decoration would form a basis. Each case has to be decided on its own facts and if the contractor's setting-out error really was impossible to live with, demolition and rebuilding could be ordered.

In all cases where errors are accepted by the architect and employer, the amount to be deducted is likely to be a source of argument. It should be noted, however, that once the architect has issued his instruction accepting errors in setting-out on the basis that a deduction is to be made later, the contractor is freed from the threat of expensive corrective measures.

Since the contract is remarkably vague regarding the exact method of calculating what is to be deducted, an astute contractor should be able to keep such deductions within reasonable limits. If the contractor is unhappy about the deduction and the architect refuses to adjust it, the remedy is to refer the matter to the appropriate dispute resolution procedures. Such problems should be capable of quick resolution by adjudication.

A contractor who makes an error in setting-out is often understandably annoyed at the prospect of amending mistakes. He may unjustly accuse the architect or clerk of works of being aware of the error long before the contractor noticed it. The architect has no general duty to the contractor to detect errors [4].

Inadequate drawings

What if the architect's drawings are inadequate? The architect must determine the levels and he must provide accurate drawings [5]. The most common failure in this area is that the drawings do not indicate beyond doubt how the building is to be set out. Triangulation from several fixed points is the only certain way of avoiding misunderstandings. It is good practice for the architect to provide special setting-out drawings, shorn of all surplus information and indicating only the important setting-out dimensions.

If the drawings do not contain all relevant information, the contractor should immediately write to the architect requesting the missing details. If the architect does not supply them on time, the contractor has a claim for extension of time and loss and/or expense. The architect cannot escape from his obligations by asking the contractor to set out as he (the contractor) sees fit pending inspection by the architect. Although the contractor's course of action is, in fact, inaction until he receives proper drawings, most contractors would try to make progress by setting out a number of pegs for the architect to inspect. If the contractor decides to do this, then as soon as the architect has approved the setting-out, the contractor must produce a drawing showing the actual triangulated dimensions as set out on site and send it to the architect requesting confirmation that the setting-out is correct. If the contractor does not protect his interests in this way, he is laying the groundwork for disputes later. A suitable letter could be somewhat as follows:

To the architect

Dear Sir,

We refer to our letter of the (date) *in which we informed you that the information on your drawings was insufficient to enable us to set out the Works accurately.*

You responded by telephone, asking us to set out the Works to the best of our ability based on the information provided.

We have carried out your instructions and you visited site on (date) *and approved our setting-out.*

We enclose our drawing number (insert) *showing the principal dimensions of our setting-out and we should be pleased to receive your written confirmation of approval before we proceed.*

If we do not receive such confirmation by return of post, we shall be obliged to notify you of delay to the Works and disruption for which we will seek appropriate financial recompense.

Yours faithfully,

The contractor will be involved in a considerable amount of additional work if he elects to attempt to set out from inadequate drawings and this procedure cannot be recommended. It would not be so bad if the contractor could claim payment for his additional work, but an instruction from the architect to the contractor to, in effect, design the siting of the building is not an instruction empowered by the contract (see section 4.2).

It cannot be stressed too strongly that many contractors are the authors of their own misfortune. This is not because they are trying to be devious, but is quite the reverse: they are trying to maintain progress. In so doing, they often shoulder more responsibility than the contract would impose. A large part of the problem is that the contractor cannot afford delays. Despite what some architects may think, a contractor never gains by delay, even if he escapes liquidated damages and recovers loss and/or expense.

The contractor's ideal is to complete the Works in less than the contract period. Thus, if he simply sits and waits for the full setting-out information, he is losing money. He may be successful in claiming some or all of his losses at some time in the future, but that does not help the contractor to fund the work at the time he most needs it. It is therefore likely that contractors will continue to try to overcome deficiencies in drawings, particularly setting-out drawings, for no reward. An awareness of the architect's obligations, however, can make the contractor's lot easier.

4.2 Release of information and architect's instructions

The introduction of an information release schedule applies only to JCT 98 and IFC 98. Whether or not one is to be provided should be stated in the recital. If the schedule is provided, it should be noted that it is not a contract document. The provision is contained in clauses 5.4 and 1.7 respectively. The information – presumably, drawings details, schedules and so on – are to be released by the architect in accordance with the schedule. There is provision for the parties to agree that the information can be released at some other time than shown on the schedule. There is also a saving clause, so far as the architect is concerned, if he is prevented by the action or default of the contractor.

If the schedule is not provided or if it does not contain all the information which will be necessary, the architect must provide further drawings and details reasonably necessary to explain and amplify the contract drawings. He must issue this information and any instructions to enable the contractor to carry out and complete the Works in accordance with the contract. One of the important provisions of the contract is the requirement that the contractor must complete the Works by the date for completion. Therefore, the architect's obligation is to provide further drawings, information and instructions at such a time so as to enable the contractor to complete by the completion date.

The clause, however, proceeds to amplify the situation. One might say that JCT is here guilty of gilding the lily, because had the clause been left at this point, the meaning would have been relatively clear. However, it goes on to say that the architect must have regard to the progress of the Works or, if the architect thinks that practical completion is likely to be achieved before the contract completion date, he should have regard to the completion date. Otherwise, he must take into account when he believes it was reasonably necessary for the contractor to receive further drawings, details, etc. This appears to open the door to the architect being able to delay issuing further information and simply to match the contractor's progress, even if the contractor is in delay. The danger here is that a 'chicken and egg' situation will arise so that it may be difficult, with hindsight, to judge whether the architect's late issue of the information is responsible for the contractor's slow progress or vice versa.

If the contractor knows *and* has reasonable grounds to believe that the architect is not aware when the contractor should have a drawing, the contractor must advise the architect sufficiently before the time when the contractor would need the information. He is to do this to the extent that it is reasonably practicable. A master's degree in crystal-ball gazing is probably required for this exercise and no doubt contractors will continue to set out a list of when and what information is required at the beginning of the contract and update it as the work proceeds. The extension of time and loss and expense clauses have been modified and simplified accordingly.

To some contractors, the position with regard to architect's instructions is simplicity itself. If the architect issues an instruction, the contractor must carry it out and eventually the contractor will be paid for his trouble. Although this is generally true, it is by no means the whole story and sometimes it can be quite false. Provision is made for architect's instructions in JCT 98 clause 4, IFC 98 clause 3.5 and MW 98 clause 3.5.

Leaving MW 98 aside for a moment, the contractor must forthwith comply with any instruction given to him by the architect provided that the instruction is empowered by the conditions, that is to say the other clauses of the contract. 'Forthwith' in this context does not mean 'immediately'; it means 'any reasonable time thereafter' [6].

In order to discover what instructions are empowered, it is necessary to search through the whole contract to find the appropriate clauses. There are 27 such clauses in JCT 98 and 16 in IFC 98. It is important to know what instructions are empowered because, if the contractor simply carries out an instruction which the architect is not empowered to give, the contractor will be in breach of contract and the employer will have no liability to pay. In such cases, the contractor may have recourse against the architect personally.

The architect is in the position of an agent with limited authority by virtue of his own contract with the employer. The contractor need not worry what the precise terms of that authority are in any particular case. If an instruction is empowered by the contract, the contractor may – indeed must – carry it out even though the architect may not, in fact, have the employer's consent.

A good example is the situation where an architect issues an instruction for the carrying out of additional work. The architect's agreement with his client will probably stipulate that the architect must obtain the client's consent before issuing an instruction of that nature. But if the architect issues such an instruction, it is empowered by the contract and the contractor is entitled to be paid for it. If the architect has not obtained his client's consent, that is a matter between architect and client and no concern of the contractor, who is entitled to rely only on the contract between contractor and employer.

There are two exceptions to the requirement that the contractor must comply forthwith. The first concerns the issue of instructions requiring a variation of an obligation or restriction imposed by the employer in the bills or specification in working space, access to site, working hours or the execution of the work in a specific order. In such cases, the contractor need not comply provided that he makes reasonable objection in writing to the architect. A reasonable objection might well be, for example, that the contractor would be quite unable to make satisfactory progress under the new conditions. If the architect agrees, he may withdraw the instruction, but any dispute may be referred to immediate adjudication, arbitration or litigation as appropriate.

The second exception arises if the contractor is unsure whether an instruction is empowered by the contract. He may request the architect to specify the clause. The architect must specify the clause in writing forthwith. On receipt of the architect's notification, the contractor may either immediately invoke the dispute resolution procedures on the point or accept the clause nominated by the architect and carry out the instruction. If the contractor elects to carry out the instruction, it is 'deemed' for the purposes of (the) contract to have been empowered even if it is not in fact [7]. This is a useful protection for the contractor and ensures that, after such inquiry and response, the contractor is entitled to the benefits of other contract provisions in respect of the instruction – for example, the valuation of variations, extensions of time and loss and/or expense – even if the architect has made a mistake.

Whether the contractor is entitled to refuse to comply with any instructions in the exceptions mentioned pending the result of arbitration is a difficult question. Whether this is a good idea will depend upon circumstances and, if in doubt, advice should always be sought. He is certainly entitled to refuse compliance if his original objection is well founded. This calls for remarkable powers of foresight because, if the result of arbitration is that the original objection was unjustified, the contractor would be in breach of contract if he had not complied. If it were thought prudent to carry out the instruction before the result of arbitration of the matter is available, it would be wise to make it clear that the instruction is being carried out without prejudice to the reference. At that stage, an appropriate letter will be drawn up by the contractor's contractual advisor who is dealing with the arbitration.

Compliance notice

If the architect considers that the contractor is unreasonably delaying carrying out a valid instruction, he has the power to issue a notice requiring compliance within seven days of receipt. If, after seven days, the contractor has not complied, the employer (note, not the architect) may employ others to do the work and deduct all costs (which may include additional architect's and quantity surveyor's fees, etc.) from money due or to become due to the contractor. Alternatively, the employer may recover such costs as a debt, i.e. through the courts. Usually, of course, it will be deducted. There is very little the contractor can do except carry out the instruction as soon as he receives such a notice.

If the costs involved are likely to be substantial and the contractor considers the architect's insistence on compliance within seven days unreasonable, he can always write a strongly worded letter which may convince the architect or serve as useful evidence of prevailing circumstances and intentions in any future adjudication or arbitration on the point. In practice architects are usually reluctant to bring others onto the site to do the work unless relations have deteriorated considerably, the work has little prospect of being completed or the project is almost completed.

Oral instructions

JCT 98 clause 4.3 states that all instructions must be in writing and then, illogically, states what is to happen if instructions are given orally. Basically, oral instructions are of no effect. Contractors often do, but should never, comply with them. If such an instruction is given, the contractor must confirm it in writing to the architect within seven days. The architect then has seven days in which to decide whether to dissent. If he does not dissent, the instruction takes effect from the expiry of the latter seven days. From the date of the oral instruction, anything between eight and about 15 days can elapse before the instruction becomes effective.

There is really no excuse for oral instructions, and if the contractor makes it clear, in a friendly way of course, that he will not act on oral instructions, the architect may stop giving them. JCT 98 goes on to provide, in clause 4.3.2.1, that the architect can (note, not must) confirm his own instruction within seven days and in such a case, the contractor need not confirm himself, which seems to be stating the obvious. Although how the contractor is to know that the architect will confirm on, say, day six when his own obligation is to confirm is not stated.

If neither architect nor contractor confirms within the time period stated, but the contractor carries out the instruction, the contractor has no rights in the matter. The architect may, however, confirm the instruction at any time up to the issue of the final certificate and it will be deemed to have taken effect on the date it was issued orally. The moral is clear: all instructions must be in writing. If not, the contractor must immediately confirm, wait seven days for dissent and, if none, carry it out.

Although an architect may argue that he has no alternative but to give an oral instruction if it is required immediately and he is at some distance from site, such a contention carries little weight now that facsimile machines are the norm. An instruction sent by fax is properly served in writing [8].

IFC 98 makes no provision for oral instructions. They are not mentioned. Therefore, confirmation by the contractor will have no effect. Confirmation by the architect will have the same effect as if the instruction had been issued at the time of confirmation.

An architect who relies upon the absence of a written confirmation in order to avoid authorising payment later may have a nasty shock in store. Looked at in terms of what the contract says, the position is clear. Work done which is not properly instructed under the terms of the contract is work which is 'not in accordance' with the contract and the architect has a duty to order such work to be removed from site. If he does not so order, he is in breach of his contractual duty to the employer. Moreover, the contractor is in breach of his contract if he does not remove it. It appears that a contractor who carries out work on the basis of an architect's oral instruction, which the architect later refuses to confirm, is entitled to amend the work to restore it to its previous condition in order to avoid a breach of contract. There is good authority, however, to the effect that compliance with an architect's instruction issued orally is good defence to claim for damages for such breach of contract [9].

On the other hand, a contractor who accepts an oral instruction under IFC 98 and carries it out will not be able to argue for any reason that it was not empowered on the contract [10].

MW 98

The position under MW 98 is somewhat different to that under the other two forms. Instructions must be in writing. If oral instructions are issued, they must be confirmed by the architect within two days. There is no provision for confirmation by the contractor. The contractor must carry out architect's instructions forthwith and there is provision for the architect to issue a seven-day compliance notice and for the employer to engage others to do the work if the contractor defaults. The actual clause may appear to be all-embracing in its effect because it states that the architect 'may issue written instructions'. There is no mention of the instructions being empowered by the conditions (as in JCT 98 or IFC 98).

It may seem that the architect can issue instructions about any matter he sees fit. However, the architect must act within the scope of his authority. The courts usually take a dim view of provisions in contracts which appear to give one party unlimited powers and they construe them narrowly, having in mind the main object of the contract and limiting the provisions accordingly [11].

This clause only gives the architect power to issue instructions regarding the Works included in the contract. Five other clauses relate to specific instructions:

- not to make good defects at his own cost (2.5);
- to exclude persons from the Works (3.4);
- to order variations (3.6);
- to expend provisional sums (3.7);
- to reinstate and make good after loss by fire, etc. (6.3B).

In addition, the architect probably has power to instruct postponement, correction of inconsistencies, opening-up and testing, and removal of defective work, but it is by no means certain.

Issue of instructions

The issue of instructions, unless omitting work, will provide grounds for extension of time under all three contracts. In addition, under JCT 98 and IFC 98, the issue or late issue, if requested by the contractor in reasonable time, will be grounds for a financial claim under clauses 26 (JCT 98) and 4.11 (IFC 98). Under the provisions of MW 98 it is stipulated that when the architect values an instruction under clause 3.6 the valuation must include any loss and/or expense incurred by the contractor in complying with the instruction, under that form alone, and the contractor need make no special application. It is the architect's duty to include the loss and/or expense.

There are one or two popular misconceptions about instructions. Only the architect has power to issue an instruction and, as noted above, his power is limited. Neither the clerk of works nor any other consultant may issue instructions although, in practice, they often do. If the employer attempts to issue an instruction, he is in breach of contract. Alternatively, the employer may be considered to be attempting to negotiate a new contract.

If the contractor carries out an instruction issued by anyone other than the architect, he does so at his peril. He is not entitled to payment and he may be considered to be in breach of contract himself by so doing. For example, if a contractor carries out an instruction, given by the heating consultant, to install an extra radiator, the architect can require the extra radiator removed from site as being not in accordance with the contract and the contractor must bear the whole cost (see the discussion above on the confirmation of oral instructions).

What is an instruction?

The golden rule for contractors is clear. They should not carry out instructions unless they are:

- given by the architect in writing or otherwise confirmed;
- empowered by the contract;
- identifiable as instructions.

This raises the question, what is an instruction? The RIBA and ACA have published standard instruction forms which most architects use. The forms are boldly headed Architect's Instruction. It is good practice to issue all instructions on such forms; however, it is not the form but the substance which determines whether it is an instruction, and an instruction can be issued as a letter.

Sometimes the architect may simply send the contractor a drawing and a compliments slip. It is dangerous to treat this as an instruction unless there is some message on the compliments slip saying that the drawing is to be used on site. Anything sent with a compliments slip should be treated with suspicion. If the architect sends a copy of the employer's letter asking that something be done, it is not an instruction, but an invitation to the contractor to do that something at his own cost.

An item in the minutes of a site meeting probably becomes an instruction when the minutes are agreed as accurate at a subsequent meeting and possibly before that if the architect is responsible for producing the minutes. One architect regularly scribbled instructions on site on pieces of plywood and the backs of roofing tiles. Such conduct is rather precious, but the instructions are valid, provided they are signed and dated. The production of copies may be a problem.

Not all instructions imply payment. Many instructions may be simply clarifications or be issued under some provision which has no financial implication, such as JCT 98 clause 8.4 requiring the removal from site of work not in accordance with the contract.

An instruction requiring a variation is not valid, and there is no liability on the employer to pay if it is issued in respect of something which the contractor must do anyway in carrying out his obligations under the contract [12]. In passing, it may be noted that an instruction signed in the architect's name by another person is quite valid, provided that the person had the architect's authority to sign [13].

4.3 Clerk of works

A good clerk of works can make a tremendous difference to work on site in terms of defects, efficiency and general atmosphere.

Both JCT 98 and IFC 98 make provision for the appointment of a clerk of works. MW 98 makes no such provision, probably because projects executed under this contract are expected to be relatively small and uncomplicated. There is no reason, however, why provision for a clerk of works should not be made in the specification.

JCT 98 clause 12 states that the employer may appoint a clerk of works who is to be under the direction of the architect. The only duty of the clerk of works is to inspect and the contractor must give him every reasonable facility. In practice, that means that the contractor should not hinder the clerk of works and he must be allowed access to all parts of the Works. The contractor cannot be expected to go to great lengths to erect scaffolding to enable him to inspect the building, but if the scaffolding is already in position, the clerk of works must be able to use it.

Directions

If the clerk of works gives a direction to the contractor, it is of no effect. This means that not only can the contractor ignore it but also that he is in breach of contract if he takes any action, or becomes inactive, as the case may be, on account of the direction. The clause then contains the curious provision that a direction of the clerk of works will be effective if given on a matter on which the architect is empowered by the contract to issue instructions and if the direction is confirmed by the architect within two working days of issue of the direction. The direction is then deemed to be an architect's instruction effective from the date of the architect's confirmation.

Two points arise from this provision. First, the architect's confirmation must be given within two working days – an action which rarely happens. If confirmation is delayed, the direction is not deemed to be an architect's instruction. But, does it matter since the confirmation itself presumably ranks as an instruction anyway? Second, even if confirmed within two days, the direction is effective only from the date of confirmation. There seems little point in the clerk of works giving a direction in the first place. It would seem more appropriate if he simply gave the architect a telephone call and asked him to issue an instruction about the matter.

The usual scenario is that the clerk of works issues a direction. The contractor need not, in fact must not, take any action on it no matter how urgent. The contractor may feel certain that it will be confirmed and either does the work and takes a chance or possibly holds back from that portion of the work until he hears from the architect. If the contractor takes the latter course, he may suffer disruption and/or delay for which he has no redress.

If he takes the former course, the architect may not confirm and the contractor will not be paid for any additional or substituted work. Worse, the architect may issue instructions, in accordance with clause 8.4, that the work done in response to the direction of the clerk of works is not in accordance with the contract and must be removed from site. The moral is quite clear: if given a direction by the clerk of works, the contractor should ignore it, carry on working as usual and if confirmation is received from the architect then, and only then, carry out the direction. If, as a result, the work is disrupted, the contractor will be able to make application for loss and/or expense.

Clearly, this is not the way a contractor usually works. Directions from the clerk of works are quite often treated as though they are architect's instructions from the moment of issue.

IFC 98 clause 3.10 is much shorter. It merely makes reference to the clerk of works being appointed by the employer as an inspector under the direction of the architect. The clerk of works is not empowered to give directions, even directions which may be ignored! From this point of view IFC 98 appears eminently sensible. Some clerks of works have expressed the view that they should be empowered to issue instructions under the contract. Such provision would obviously lead to confusion on site.

The duty of the clerk of works is to the employer. He has no duty towards the contractor to find defects. The approval of the clerk of works counts for nothing. It may or may not be indicative of the architect's attitude. If the clerk of works points out defects, the contractor would be wise to take notice, but there is no substitute for a competent site agent or, as the contract says, person-in-charge.

The contractor will often present daywork sheets to the clerk of works for signing. This is incorrect practice. JCT 98 clause 13.5.4 stipulates that such vouchers must be delivered to the architect or his authorised representative for verification before the end of the week following the week in which the work has been carried out. The clerk of works is not the architect's representative unless the architect specifically states the same in writing to the contractor.

A negligent clerk of works will not remove liability from the architect so far as the employer is concerned, but if the employer successfully sues the architect for negligence, the damages payable by the architect may be reduced on account of the negligence of the clerk of works [14]. This is a point of more interest to the architect and the employer than to the contractor. It depends on the clerk of works being employed by the employer and not by the architect, as sometimes happens, and raises the point of the employer's vicarious liability for the actions of the clerk of works.

Defacement

Many clerks of works are in the habit of using chalk or wax crayon to deface work or materials which they consider to be defective. In this they are sometimes encouraged by the architect and the contractor himself. Some contractors, however, quite rightly take exception to such conduct and protest to the architect. If materials are defaced by the clerk of works it is because, presumably, they are not in accordance with the contract. Either they will be removed by the contractor or the architect will instruct removal. They are, therefore, the contractor's property. It may be that the contractor could use materials in other less stringent situations except for the fact that they are defaced. There is no doubt that the clerk of works is not entitled to deface Works or materials. His role is purely to inspect. The contractor should put a stop to defacement as soon as it first appears. This is best done by a suitably worded letter on the following lines to the architect:

> *Dear Sir,*
>
> *It is common practice for the clerk of works to deface work or materials which he considers to be defective. The basis for such action presumably is to bring the defect to the notice of the contractor and ensure that it cannot remain without attention.*
>
> *We object to the practice on the following grounds:*
>
> *1. The work or materials so marked may not be defective and we will be involved in extra work and the employer in extra costs in such circumstances.*
>
> *2. The work or materials so marked, if indeed defective, will not be paid for and will be our property when removed. We may be able to incorporate it in other projects where a different standard is required. Defacement by the clerk of works would prevent such re-use.*
>
> *We will take no point about the defacing marks we noted on site today, but if the practice continues, we will seek financial reimbursement.*
>
> *Yours faithfully,*

Specialist clerks of works

It is the practice of some large organisations, who maintain a body of clerks of works on their permanent staff, to allow a number of so-called 'specialist' clerks of works to inspect the site at regular intervals. It is not always appreciated that the contractor has possession of the site under licence from the employer and, with the exception of the persons noted in the contract, he has the right to refuse admittance to the site. The situation is clearly delicate, but the contractor would be ill-advised to allow a multitude of inspectors to roam the site. Of course, it is always open to the architect to make such 'specialist' clerks of works his authorised representatives, in which case the contractor must allow them access, certainly under JCT 98 provisions (clause 11).

The architect is unlikely to make large numbers of persons his authorised representatives because it would imply that they can act for him, issue instructions, etc. The situation is not absolutely clear, but any contractor faced with 'specialist' clerks of works might try refusing access until such time as the architect has clarified the position in writing.

Snagging

A clerk of works will often issue snagging lists, particularly as practical completion draws near. If the contractor accepts such lists as merely helpful reminders of work to be done, all should be well. Problems and disputes sometimes arise on site because the clerk of works produces snagging lists and after items receive attention the architect produces another list. Contractors often feel aggrieved about it, maintaining that one list is quite enough. In fact, the snagging list is not of contractual significance unless issued by the architect and he clearly states that it represents the only points requiring attention before he is prepared to issue a practical completion certificate. It is the contractor's obligation to complete the building in accordance with the contract and the clerk of works can never dispense him from that obligation. The clerk of works has no duty to produce snagging lists.

A clerk of works can be very useful but it should always be remembered that he is just an inspector. He is empowered to look and to note and that is all. If he issues directions, the contractor should not act on them, unless of course they relate to obvious defects, until they are confirmed. Confirmation by the contractor is not effective.

References

1. Robinson v Harman (1848) 154 ER 363.
2. Forsyth v Ruxley Electronics and Construction Ltd (1995) 73 BLR 1.
3. William Tomkinson and Sons Ltd v The Parochial Church Council of St Michael (1990) 6 Const LJ 319.
4. Oldschool v Gleeson (Construction) (1976) 4 BLR 103.

5. London Borough of Merton v Stanley Hugh Leach Ltd (1985) 32 BLR 51.
6. London Borough of Hillingdon v Cutler (1967) 2 All ER 361.
7. For an interesting judicial comment about 'deeming' see: Re Coslett (Contractors) Ltd, Clark, Administrator of Coslett (Contractors) Ltd (In Administration) v Mid Glamorgan County Council [1997] 4 All ER 115.
8. Hastie and Jenkerson v McMahon (1990) The Times 3 April 1990.
9. G Bilton & Sons v Mason (1957) unreported.
10. Bowmer and Kirkland Ltd v Wilson Bowden Properties Ltd (1996) 80 BLR 131.
11. Glyn & Others v Margetson & Co and Others (1893) AC 351.
12. Sharpe v San Paulo Brazilian Railway Co (1873) 8 Ch App 597.
13. Anglian Water Authority v RDL Contracting Ltd (1992) 27 Con LR 76.
14. Kensington and Chelsea and Westminster Area Health Authority v Wettern Composites (1984) 1 Con LR 114.

5 Money

5.1 Certificates

JCT 98 provides for the issue of 10 different kinds of certificate. The figure for IFC 98 is six and for MW 98 it is five. Many of these certificates are not financial, although most contractors associate the word 'certificate' with money. In order to be a valid certificate, a document must be headed 'Certificate', start with the words 'I certify' or be clearly referenced to the contract provision empowering issue. In any event, it must be the clear expression of the judgment, skill or opinion of the architect [1].

Financial certificates are covered by JCT 98 clause 30, IFC 98 clauses 4.2 to 4.10 and MW 98 clause 4. JCT 98 and IFC 98 provisions are similar and will be discussed together. MW 98 will be considered separately. What follows can be no more than a brief summary and the parties are advised to study the contract terms carefully.

Interim certificates are to be issued at the dates noted in the Appendix, usually at monthly intervals. They must state the amount due to the contractor from the employer and the employer has 14 days in which to pay from the date of the certificate. Failure to honour certificates within the due time is a ground for suspension or determination by the contractor (see section 7.2) and entitles the contractor to interest. If the architect so wishes, he may ask the quantity surveyor to carry out a valuation not more than seven days before the date of the certificate, but responsibility for the correctness of the sum on the certificate remains with the architect [2]. Alternatively, the contractor may submit an application for payment not later than seven days before certification date which the quantity surveyor may either accept or reject by sending the contractor the quantity surveyor's own view in the same detail as submitted by the contractor. The amounts to be included in each certificate are as follows. Subject to retention:

- total value of work properly executed;
- total value of materials delivered to site for incorporation, if not premature;
- value of listed off-site materials;

- certain payments to nominated sub-contractors (JCT 98 only);
- certain profits in respect of nominated sub-contractors (JCT 98 only).

Not subject to retention:

- in respect of payments or costs due in regard to opening-up and testing, loss and/or expense, statutory obligations, provisional sum insurance, insurance premiums;
- final payments to nominated sub-contractors (JCT 98 only);
- fluctuation payments;
- certain payments to nominated sub-contractors (JCT 98).

Less any authorised deductions.

The certificate should express the amount due to the contractor as the balance between the total of the above amounts and the amounts already certified on previous certificates.

Priced activity schedule

Priced activity schedules apply to JCT 98 (clause 30.2.1.1) and IFC 98 (clause 4.2.1(a)) if the Appendix states so. The activity schedule must be attached to the Appendix with each activity priced so that the total of the prices equals the contract sum, less provisional sums, prime cost sums, contractor's profits on this and the value of any work for which approximate quantities are included in the bills of quantities. The contractor has an option whether or not he wishes to provide an activity schedule. If he does, it is used to assist in the valuation of interim certificates. The prices in the activity schedule are to be proportioned so as to give the amount in the interim certificate. This is likely to mean that the contractor will receive sums in the interim certificates which more closely reflect the work carried out.

A priced activity schedule is usually used as an alternative to priced bills of quantities. It seems strange, at first sight, to use it as well as bills of quantities. It should be noted, however, that even where a priced activity schedule is used, bills of quantities are still used for the valuation of variations.

Advance payment

Both JCT 98 and IFC 98 make provision for advance payment, except for local authorities. The employer can pay a sum to the contractor on a date stated in the Appendix. The idea is that the sum will be paid early, in order to assist the contractor to finance the project. He must reimburse the employer the amounts, and at the intervals, which the two parties agree and state in the Appendix. In these circumstances, the contractor is required to give a bond, usually in the form reproduced at the back of the contract (JCT 98 clause 30.1.1.6 and IFC 98 clause 4.2(b)).

Retention

A percentage, usually five per cent, is held by the employer, who may use the money whenever the contract directs that he may deduct from money due or to become due to the contractor. This retention fund is also useful at the end of the job to ensure that making-good of defects is carried out. The retention money is held on trust for the contractor. That means that, although it is held by the employer, it really belongs to the contractor, which is why the employer has only limited powers of access to it. It is settled that the contractor has the right to demand that the retention money be paid into a separate bank account, clearly named in favour of the contractor [3].

Indeed, private editions of JCT 98 include a provision to that effect. The importance of keeping trust money separate is that, if the employer becomes insolvent, clearly identified trust money can be paid to the contractor. Otherwise, he has to take his chance along with other creditors. The employer has no right to use the money for his own purposes and to do so would be a breach of trust. The employer has an obligation to keep the trust money in a separate bank account even if there is no clause to that effect or if such a clause has been struck out [4]. The contractor is not obliged to make a request for the money to be placed in a separate account each time a certificate is issued and paid. He may make just one request at any time [5]. If the employer goes into receivership before a separate account has been set up, it seems that the contractor has lost his right to the money [6].

The contract states that the employer has no obligation to invest the money. In other words, the contractor is not entitled to interest. Since there is a statutory obligation to invest, which a contract provision cannot defeat, it is possible that a contractor who insisted would be entitled to interest. Note, however, that no contractor has, so far, felt confident enough to test the point through the courts.

The whole of the retention clause in IFC 98 (clause 4.4) is stated to apply only where the employer is not a local authority. Therefore, if the employer is a local authority, the position is not clear and the clause should be amended. Half the retention must be released to the contractor when the practical completion certificate is issued. The remaining half is released, under JCT 98 terms, on the issue of the certificate of making good defects. Under IFC 98 terms the remaining half retention is not released until the final certificate.

Final certificate

Not later than six months after practical completion, the contractor must send to the architect, or to the quantity surveyor if so directed, all documents necessary for working out the final contract sum. Not later than three months thereafter, the quantity surveyor must prepare a statement of all valuations of variations and send the contractor a copy of the computations of the adjusted contract sum. JCT 98 provides detailed guidance on the way in which the contract sum is to be adjusted. In the case of JCT 98, copies of the final computation must be supplied to the nominated sub-contractors.

The rules for the issue of the final certificate vary between JCT 98 and IFC 98. In the case of JCT 98 the final certificate must be issued not later than two months from the latest of the following:

- end of the defects liability period;
- issue of certificate of making good defects;
- receipt by architect or quantity surveyor of all the documents necessary for carrying out the final computation.

It is common experience for the last event to be the determining factor.

In the case of IFC 98, issue must take place within 28 days of the sending to the contractor of the computations of the adjusted contract sum or the issue of the certificate of making good defects, whichever is the later.

The final certificate must state the adjusted contract sum, the amounts previously certified and the balance expressed either as a sum due to the employer or a sum due to the contractor. The employer or contractor, as the case may be, must pay within 28 days of the date of issue.

Unless adjudication, arbitration or other proceedings have been commenced by either party within 28 days of the date of issue, the final certificate is conclusive evidence that:

- where quality or standards of materials, goods or workmanship are expressly stated to be a matter for the architect's satisfaction, he is satisfied;
- the contract terms requiring adjustment of the contract sum have been correctly applied;
- all due extensions of time have been given;
- reimbursement of loss and/or expense is in final settlement of all matters under clause 26 (IFC 98 clause 4.12).

The only exceptions arise in the case of fraud or accidental inclusions or exclusions of work, materials or figures in any computation or arithmetical error. It is clearly stated that no other certificate is conclusive evidence that work or materials are in accordance with the contract. The current position so far as the first item is concerned has been discussed in section 1.2.

Not less than 28 days before the issue of the final certificate, JCT 98 provides that the architect must issue an interim certificate including all the finally adjusted sub-contract sums. Throughout the certifying process under this contract, the architect is required to direct the contractor regarding the amounts included from time to time in favour of nominated sub-contractors and to inform the sub-contractors accordingly.

MW 98 provisions stipulate that the architect must certify progress payments to the contractor at four-weekly intervals. It is possible that the parties may come to some other arrangements in the case of a small job, such as stage payments. The employer must pay within 14 days of the date of issue of the certificate. Each certificate must state the value of Works properly executed less amounts previously certified. The retention, which is not expressed as being held in trust, is normally five per cent. If the employer becomes insolvent, the contractor's retained money is not protected and the contractor is not entitled to demand that the money be kept in a separate bank account. Half the retention is to be released within 14 days of the certificate of practical completion.

The contractor must supply within the specified period, usually three months, from practical completion all documentation for computation of the final certificate. The final certificate must be issued within 28 days of receipt of the documentation, provided that a certificate of making good defects has been issued. The final certificate is not stated to be conclusive as regards any matter, not even the amount.

These contracts are lump sum contracts. The contractor is entitled to payment provided he completes substantially the whole of the work [7]. The fact that interim payments are made does not alter the position. Failure to perform substantially means that the contractor cannot recover [8].

Reference in the contracts to work 'properly executed' refers to work which is in accordance with the contract. The contractor is not, of course, entitled to be paid for defective work. If the architect does certify defective work, he may find himself personally liable if the contractor becomes insolvent [9].

Off-site materials

It should be particularly noted, that if the employer wishes to pay for materials off site, he will have listed the materials and attached the list to the contract bills of quantities. The items on the list must be divided into two categories: the first category is 'uniquely identified items' and the second category is 'not uniquely identified items'. Uniquely identified items are those materials or goods which are easy to recognise, such as heating boilers or sanitary fittings or the like. Items not uniquely identified cover such items as bricks, sand, tiles, timber and anything which it would be difficult to recognise as belonging to a particular site.

The architect is obliged to include all such listed materials and goods in the valuation in any certificate and he must include them before delivery to site. This is subject to the contractor fulfilling certain specified criteria. They are as follows:

- The contractor must have provided the architect with reasonable proof that he owns the items, so that after he has been paid the value of items included in certificates, the ownership of items will pass to the employer.
- The contractor must have provided the employer with a bond if so required by an entry in the Appendix. The surety for the bond must be approved by the employer and, unless otherwise agreed, the terms of the bond must be as agreed between the JCT and the British Bankers Association. In the case of not uniquely identified materials, the provision of a bond is mandatory.
- The items must be in accordance with the contract.
- They must be kept at the premises where they have been manufactured or stored and they must either be set on one side or, alternatively, they must be clearly marked so as to identify the employer and that they are destined for the Works.
- The contractor must have provided reasonable proof that the items are insured in respect of specified perils.

There is no provision for payment for off-site materials under MW 98.

Set-off

The question of set-off has long been a bone of contention. In general, it appears that there is a right of set-off unless, looking at the contract as a whole, it is excluded expressly or by necessary implication [10]. The point is important because, at one time, the architect's certificate was considered as good as cash and must be honoured [11]. It now seems clear, however, that provided the employer has good grounds he can withhold payment on a certificate and, resisting summary judgment, go to arbitration or trial [12]. This is bad news for contractors, particularly because it appears that the employer need only show that there are reasonable grounds to challenge the certificate [13]. Following the coming into force of the Arbitration Act 1996, it will be rare for a contractor to obtain summary judgment if the employer fails to pay and the contractor may be wiser to use the contractual power to determine their employment.

Following the Housing Grants, Construction and Regeneration Act 1996, all three contracts contain express provisions to deal with set-off or, as the contracts now refer to it, withholding or deduction (JCT 98 clauses 30.1.1.3, 30.1.1.4, 30.1.1.5, 30.8.2, 30.8.3 and 30.8.4; IFC 98 clauses 4.2.3(a) and (b), 4.3(b) and (c), 4.6.1.2 and 4.6.1.3; MW 98 clauses 4.4.1,

4.4.2, 4.5.1.2 and 4.5.1.3). Because the provisions were inserted as a result of legislation, they are the same in each contract.

The employer must issue a written notice to the contractor within five days of the date of issue of each certificate (including the final certificate). The notice must state the amount which the employer proposes to pay, to what it relates and how it is calculated. Presumably, the notice will state the amount in the certificate although it is possible, but not certain, that the employer could use the opportunity to abate the sum (that is, to reduce it, possibly because work has not been done). If the employer wishes to withhold or deduct any amount from the sum due, he must issue a written notice not later than five days before the final date for payment of any certificate (again, including the final certificate). This notice must state the grounds for withholding and the amount to be withheld for each ground. If the employer does not give a written notice in accordance with one and/or both provisions, he must pay the certified sum in full. Under the Act the first notice will suffice for both if it indicates a deduction and sets out the grounds in sufficient detail. It appears from the wording that the contractual position may be the same, but the wise employer will always give both notices if wishing to deduct any amounts. Obviously, the employer who intends to pay the full amount may omit to give any notices. He would technically be in breach but, provided he paid in full, the contractor could hardly complain.

JCT 98 clauses 30.1.1.1, 30.8.5, IFC 98 clauses 4.2(a) and 4.6.2 and MW 98 clauses 4.2.2 and 4.5.2 provide that the employer must pay simple interest at five per cent above Bank of England Base Rate if he fails to pay the amount due by the final date for payment. This is in addition to the contractor's other rights to suspend or determine or, in appropriate cases, to accept repudiation under the general law.

What if the architect simply refuses to issue a certificate? If the contractor can show that the refusal is due to interference by the employer, he is entitled to determine his employment. Under JCT standard form contracts the contractor cannot recover money without a certificate. That is because where there is an arbitration clause the courts take the view that the contractor's remedy for the absence of a certificate is to seek arbitration [14]. In the rare instance of a building contract not containing an arbitration clause, it is likely that the contractor could sue without a certificate [15]. In general, the architect's failure to issue a certificate at the proper time is a breach of contract for which the employer is liable [16].

If the contractor can show that the architect deliberately interfered with the contract in order to keep the contractor out of money, the architect may be personally liable to the contractor for the tort of interference with contractual rights [17]. It is not something which is easily proved.

It is often thought that the architect named in the contract must sign all certificates for them to be valid. That is not correct. It is quite sufficient if a properly authorised person signs in the name of the named architect [18]. It matters not that the person is not an architect, provided the person is authorised by the architect to sign in his name. It should be noted, however, that the architect's signature on a certificate does not amount to issuing the certificate. To accomplish that under the provisions of these contracts the architect must at least bring the contents of the certificates to the attention of the employer [19].

5.2 Variations

Contractors sometimes ask if they are obliged to carry out an architect's instruction requiring a variation. The answer is 'yes' as far as JCT contracts are concerned, but there are some points to note. The clauses authorising the architect to issue instructions requiring variations are JCT 98 clause 13, IFC 98 clause 3.6 and MW 98 clause 3.6. In the absence of these clauses, the architect would have no power to order variations and any attempt to do so by architect or employer would amount to a breach of contract and the contractor may be justified in negotiating a new price for the contract. This is because JCT 98, IFC 98 and MW 98 are lump sum contracts and, basically, the contractor has tendered a price for the whole of the work.

There are some other clauses which authorise variations in specific circumstances (JCT 98 clause 2.2.2.2, IFC 98 clause 5.4.3, etc.), but the principal clauses are the ones noted above. JCT 98 and IFC 98 provisions are similar and will be considered first. Variations are defined in these contracts as the alteration or modification of the design or the quality or quantity of the Works from that shown in the contract documents. They include: additions, omissions and substitutions; alterations of kinds or standards of materials; or the removal from site of materials delivered for the Works, unless the removal is because the materials are defective. That is all very clear and gives the architect wide powers.

An important provision is the architect's power to issue instructions imposing, adding to, omitting from or altering obligations or restrictions relating to access, working space, working hours or the order of the work. This seems to be a dangerous extension of the architect's powers into the contractor's own field, and there is only one safeguard: the contractor can make reasonable objection under JCT 98 clause 4.1.1 or IFC 98 clause 3.5.1.

The architect may not omit work, which has been measured in the bills and priced by the contractor, in order to give it to others to carry out [20]. The prohibition also covers the omission of provisional sums to allow others to do the work [21].

Valuation

Variations are to be valued by the quantity surveyor in accordance with the rules laid down in the contract. This procedure may be set aside if the contractor and the employer (not the architect) agree. This allows the employer, for example, to accept the contractor's quotation for a proposed variation. Alternatively, under JCT 98, there is provision for a contractor's quotation, either requested or unrequested. In IFC 98 there is simply provision for an unrequested quotation. The effect of this is discussed later in this section. The rules for the valuation of variations are sensible. In JCT 98 reference is to be made to the prices in the priced bills. In IFC 98 reference is to be made to the prices in the priced document (see section 1.1).

Omissions are to be valued at the rate in the bills. Additions and substitutions which are similar in character and executed under similar conditions and without significant change in quantity to work in the bills are to be priced using those rates. If there are changes in the conditions and quantity, valuations are to use the bill rates as a basis. If the work is not of a similar character or involves other than additions, omissions or substitutions or if it is not reasonable to value it using bill rates as a basis, then a fair valuation must be made. Rules are set out for the valuation of work on a daywork basis if it cannot properly be measured.

The above is only a general summary and all parties are advised to study the relevant clauses carefully. It should be remembered, however, that the quantity surveyor has a fair degree of discretion regarding the method of valuation and he is under no obligation to obtain the contractor's agreement. The amount to be included in any certificate in respect of such valuation is for the architect to decide [22] even though, in practice, the architect will usually adopt the quantity surveyor's valuation. Even in adopting the quantity surveyor's valuation, however, the architect has a duty to his client to carry out sufficient checks to satisfy himself that the certified sum is correct. That is not to say that the architect is obliged to revalue the work himself – that would be needless duplication – but the architect should ask for sufficient supporting information from the quantity surveyor so that he can easily see what work has been valued. The rules for valuation, where approximate quantities are involved, and the implications of defined and undefined provisional sums should be carefully studied where Standard Method of Measurement 7th edition (SMM 7) is to apply.

Important provisions require that if an instruction causes substantial changes in the conditions under which other work is carried out, such other work will be treated as if varied. For example, if the architect issues an instruction changing 'dry lining' to 'three coats wet plaster work', the contractor will be entitled to be paid for that variation as appropriate. By virtue of JCT 98 clause 13.5.5 or IFC 98 clause 3.7.8, he will also be entitled to additional payment because altered conditions may lead him to carry out some of his other operations in a different order or may necessitate added protection. The change in conditions must be substantial.

The variations clauses do not seem to include valuation of the effect of a variation upon the regular progress of the work. Any claims under this head must be made in accordance with JCT 98 clause 26 or IFC 98 clause 4.11. The words are: 'No allowance shall be made (in the valuation) for any effect upon the regular progress of the Works or for any other direct loss and/or expense for which the contractor would be reimbursed by payment under any other provision in the conditions.' If, however, the contractor can show that he would not be so reimbursed under another provision, he is entitled to have the amount valued under the variation clause.

There is an alternative valuation procedure in JCT 98. It allows the architect to instruct the contractor that a variation is to be dealt with under clause 13A. The contractor has 21 days from receipt of the instruction or requested further information to submit his quotation to the quantity surveyor. The quotation must include the value of the work, any extension of time, loss and/or expense and the cost of preparing the quotation. The architect may request other information such as a method statement. Acceptance must be within seven days from receipt and be confirmed by the architect. If not accepted, valuation is to be carried out under the normal rules or the architect must instruct that the variation is not to be carried out. The great advantage is that otherwise contentious claim matters are settled at the time of the variation. Points for contractors to watch:

- The contractor has seven days after receipt of the instruction to give written notice of disagreement with the application of clause 13A.
- If a later variation is instructed to clause 13A work, neither the normal rules nor clause 13A apply. Instead, the quantity surveyor must make a fair and reasonable valuation based on the earlier clause 13A quotation.

The contractor may now give a quotation without being asked. It is called the 'contractor's price statement' and it is dealt with by JCT 98 clause 13.4.1.2 alternative A and IFC 98 clause 3.7.1.2 option A. The provisions are virtually identical. The contractor may, within 21 days from receiving an instruction or from commencing work in connection with an approximate quantity in the contract, give the quantity surveyor a price for carrying out the work. In addition he may, if he wishes, include what he requires so far as extension of time is concerned and any loss and/or expense. The price statement must be calculated by the contractor in accordance with the valuation rules in the contract.

The quantity surveyor has 21 days from receiving the price statement to tell the contractor in writing either the statement is accepted or that it is not accepted. If it has been accepted, the amount is to be added – or deducted, of course, as appropriate – from the contract sum. If not accepted, the quantity surveyor must tell the contractor why it is not accepted and the contractor must be given the reasons in the same detail as given by the contractor in its price statement. The quantity surveyor must also supply an amended price statement. The contractor has 14 days from receipt of the amended statement to decide whether he wishes to accept it. If it is not accepted, either party may refer the contractor's price statement and the amended price statement to adjudication.

Where a contractor has included his extension of time and loss and/or expense requirements, the quantity surveyor has 21 days in which to notify the contractor that they are acceptable or that the appropriate clauses in the contract will apply.

It is noteworthy that before the quantity surveyor comes to his decisions, he must consult the architect. However, it appears that the quantity surveyor could make a decision in the face of opposition from the architect. The strange situation of a quantity surveyor agreeing an extension of time to which the architect dissents has already arisen on more than one occasion, to the author's knowledge. Clearly, an extension of time agreed under these provisions does not require the architect to ratify it under JCT 98 clause 25 or IFC 98 clause 2.3.

MW 98

Under MW 98 variations are empowered by clause 3.6. It is a very short clause allowing the architect to vary the Works by addition, omission or change or to change the order or period in which they are to be carried out. The contractor has no right of reasonable objection, but he can always refer any dispute to arbitration.

Valuation of variations may be carried out by the architect and contractor reaching agreement before the work is carried out or by the architect, on a fair and reasonable basis using the prices in the priced specification, schedules or schedule of rates where relevant. The architect's view as to whether the particular priced document is 'relevant' in any particular instance will, doubtless, prevail, at least until adjudication. The contractor is not in a strong position and, even if priced schedules are used, the employer does not warrant that they are correct. A major difficulty, from the contractor's point of view, is that he is deemed to have included in his price for carrying out and completing the Works in accordance with the contract documents. Thus, work shown on the drawing, but not in the schedules, does not rank as a variation as would be the case under JCT 98 with Quantities edition. (See also section 6.2 in relation to the inclusion of loss and/or expense.)

MW 98 makes provision for a quantity surveyor to be appointed (4th recital), but curiously makes no further reference to him. If a quantity surveyor is appointed, the architect will probably delegate to him the carrying out of valuations.

Points

All three forms make reference to the fact that no variation will vitiate or invalidate the contract. This is superfluous, because nothing empowered by the contract can invalidate it. It must not be thought, however, that the architect can order variations with impunity. If the variation or the sum of all the variations on a particular contract is such that the whole scope and character of the work is changed, the contractor may be entitled to negotiate a totally new contract. The change must be such that the contractor can say that the project is not substantially that for which he originally tendered. For example, if virtually every detail is changed little by little so that flat roofs are replaced by pitched, baths by showers, small wooden windows by large metal windows and so on, the contractor may well have a case. It should be noted that the mere number of variations is not important and each situation must be considered on its merits.

All variations, except omissions, entitle the contractor to extensions of time if they cause the completion date to be exceeded. They also entitle the contractor to be reimbursed for direct loss and/or expense.

A situation which sometimes arises concerns work which the architect insists is included in the contract, but which the contractor is equally certain should be an extra. Certain groundworks sometimes fall into this category. The architect may refuse to authorise a variation. What is the contractor to do? The work may be substantial in quantity and value. If the contractor is confident of his position, his remedy appears to be to refuse to carry out the work unless a variation is authorised and, failing such authorisation, to treat the contract as repudiated and sue for damages [23].

This may not always be a prudent action to pursue, the contractor risking the possibility of bearing huge losses if the result of legal action is not in his favour. If the contractor proceeds with the work and attempts to claim later, it may be considered that he has carried out the work on the architect's interpretation of the contract and he is entitled to no extra payment. There is no easy answer but, at the very least, if the contractor elects to proceed, he should make clear his position in writing to the employer and do the work 'without prejudice' to his rights to claim later.

In some circumstances, it may be held that the employer has implicitly promised to pay if the work is, in fact, additional to the contract [24]. In practice, it should be possible for the parties to agree that the contractor carries out the work on the clear understanding that the matter can be settled by adjudication or arbitration and payment made if the work is found to be a variation. A suitable letter might run along the following lines:

> *To the Employer*
>
> *SPECIAL DELIVERY/RECORDED DELIVERY WITHOUT PREJUDICE*
>
> *Dear Sir*
>
> *We refer to your letter of* (date) *and ours of* (date) *relating to* (describe work). *It is our firm view that this work is not included in the contract and, therefore, constitutes a variation for which we are entitled to payment.*
>
> *We are advised that we can refuse further performance until you authorise a variation. If you continue to refuse to so authorise, we may treat the contract as repudiated and sue for damages.*
>
> *Without prejudice to our rights, we are prepared to carry out the work, leaving this matter in abeyance for future determination by adjudication or arbitration, if you will agree in writing and confirm that you will not deny our entitlement to payment in such reference if, on the true construction of the contract, the work is held to be not included.*
>
> *Yours faithfully,*

Dayworks

The valuation of dayworks is often a contentious issue. If the work is to be valued by measurement or on some other basis, the submission of signed daywork sheets by the contractor will be irrelevant. It is only if the work is to be carried out and valued on a daywork basis that signed daywork sheets assume any importance. If the sheets are signed by the architect's authorised representative, they must be valued using the hours and resources on the sheets. The quantity surveyor has no power to go behind the sheets and substitute his own estimate of the hours it should have taken to do the work [25]. Signing under the phrase 'for record purposes only', which is very common, does not imply that there is power to value the work on a different basis [26]. Even where sheets have not been signed, they will be good evidence of the hours spent provided that they were completed by the contractor at the end of the relevant day.

References

1. Token Construction Co Ltd v Charlton Estates Ltd (1973) 1 BLR 50.
2. R B Burden v Swansea Corporation [1957] 3 All ER 243.
3. Rayack Construction Ltd v Lampeter Meat Co Ltd (1979) 12 BLR 30.

4. Wates Construction (London) Ltd v Franthom Property Ltd (1991) 53 BLR 23.
5. J F Finnegan Ltd v Ford Sellar Morris Developments Ltd (1991) 53 BLR 38.
6. Macjordan Construction Ltd v Brookmount Erostin Ltd (1991) CILL 704.
7. Hoenig v Isaacs [1952] 2 All ER 176.
8. Bolton v Mahadeva [1971] 2 All ER 1322.
9. Sutcliffe v Thackrah [1974] 1 All ER 319.
10. Gilbert-Ash (Northern) Ltd v Modern Engineering (Bristol) Ltd (1973) 1 BLR 73.
11. Dawneys Ltd v F G Minter Ltd (1971) 1 BLR 16.
12. C M Pillings & Co Ltd v Kent Investments Ltd (1985) 4 Con LR 1.
13. R M Douglas Construction Ltd v Bass Leisure Ltd (1991) 53 BLR 119.
14. Lubenham Fidelities & Investments Co Ltd v South Pembrokeshire District Council and Wigley Fox Partnership (1986) 6 Con LR 85.
15. Panamena Europea Navigacion (Compania Limitada) v Frederick Leyland & Co Ltd (1947) AC 428.
16. Croudace Ltd v London Borough of Lambeth (1986) 6 Con LR 70.
17. Edwin Hill & Partners v First National Finance Corporation PLC [1988] 07 EG 75.
18. London County Council v Vitamins Ltd [1955] 2 All ER 229.
19. London Borough of Camden v Thomas McInerney & Sons Ltd (1986) 9 Con LR 99.
20. Carr v J A Berriman Pty Ltd (1953) 27 ALJR 273; Commissioner for Main Roads v Reed & Stuart Pty Ltd (1974) 12 BLR 55. Vonlynn Holdings Ltd v Patrick Flaherty Contracts Ltd (1988) unreported.
21. AMEC Building Ltd v Cadmus Investments Co Ltd (1996) 13 Const LJ 50.
22. R B Burden Ltd v Swansea Corporation (1957) 3 All ER 243.
23. Peter Kiewit Sons' Company of Canada Ltd v Eakins Construction Ltd (1960) 22 DLR (2d) 465.
24. Molloy v Liebe (1910) 102 LT 616.
25. Clusky (trading as Damian Construction) v Chamberlin (1994) April BLM 6.
26. Inserco Ltd v Honeywell Control Systems Ltd (1996) unreported.

6 Claims

6.1 Extension of time

The basic idea of the extension of time and liquidated damages clauses in JCT 98, IFC 98 and MW 98 is very simple and straightforward. If the contractor does not complete the Works by the date for completion, the employer is to be paid pre-agreed damages. If, however, the contractor is delayed due to certain specified events, the contract period will be extended, thus releasing him from the obligation to pay damages for the overrun in respect of those events. In essence, that is all there is to the provision. In practice, however, the application of the clauses seems to cause problems out of all proportion to the issues at stake. Part of the difficulty lies in the number of myths and misunderstandings which cloud the sensible operation of the contract provisions.

The extension of time provisions are for the benefit of the contractor and the employer. It is easy to see that the contractor benefits from an extension of time because it releases him from the obligation to pay liquidated damages. The benefit to the employer is a little more complex.

Under JCT 98 clause 23 and IFC 98 clause 2.1 the contractor is obliged to commence the Works on the date for possession and regularly and diligently proceed with them so that they are complete by the date for completion stipulated in the contract. MW 98 clauses 1.1 and 2.1 contain terms to much the same effect.

It is important to remember that under the general law the contractor's obligation to complete the Works by the contractual completion date is removed if the employer or his agents are responsible for some or the whole of the delay [1]. Such actions as the issue of instructions or any kind of interference or obstruction fall into this category. In such cases time becomes 'at large', that is to say that there is no longer any date by which the contractor must complete and, therefore, no date from which liquidated damages can be calculated [2]. The contractor's obligation is then to complete the Works within a reasonable time. A reasonable time may, of course, be the length of the original contract period plus the period for which the employer has caused delay. The employer may still claim damages, but instead of merely being able to deduct them, he is faced with the problem of having to prove them in court.

The above general rules may be amended if there is an express term in the contract [3]. There is such an express term in the three contracts under consideration which allows the architect to grant an extension of time for employer's defaults and thus preserve the employer's right to deduct liquidated damages for any period of overrun beyond the extended date. The benefit to the employer is now clear.

The extension of time provisions in JCT 98 and IFC 98 are contained in clauses 25 and 2.3/4 respectively. (The provisions in MW 98 are somewhat different and contained in clause 2.2 which will be considered later.) The provisions clearly take account of two distinct types of delay:

- Delays caused by the employer (these are the most important).
- Delays caused by events outside the control of either the contractor or the employer.

The architect may only grant an extension of time if the event falls within the events contemplated by the extension of time clause. This means that if the employer causes delay by some default which is not included in the clause, the architect will be unable to grant an extension of time and time will become 'at large'. In deciding whether the architect has power to extend time and whether the event does fall within the events in the clause, the court will construe the terms against the person wishing to rely upon them. The clause must provide expressly or by necessary inference for an extension on account of the employer's fault [4].

There are possible faults of the employer which will not fall within the terms of JCT 98 or IFC 98 clauses and which, therefore, will cause the liquidated damages clause to be inoperative. Even more interesting is the position under MW 98 clause 2.2. Until 1988 there was no list of events for which an extension might be granted, but only the bald statement that the architect could make a reasonable extension of time for delays caused by 'reasons beyond the control of the contractor'. There are a number of authorities which considered that this clause was not specific enough to embrace any faults of the employer. If correct (the point has not, to my knowledge, been tested in the courts), any additional instructions or other delaying actions or omissions by the employer would make time 'at large'. The point appears to have been taken by the Joint Contracts Tribunal in Amendment MW 5 which added the words: 'including compliance with any instruction of the Architect under this contract whose issue is not due to the default of the Contractor'.

Amendment MW 7 added a further sentence to make clear that reasons within the control of the contractor include defaults by any person or firm employed by the contractor, e.g. sub-contractors. This was in response to a judicial decision which appeared to indicate that, in certain circumstances, the actions of a sub-contractor are beyond the contractor's control [5]. Architect's instructions are likely to be the main source of delay attributable to the employer. It is not, of course, the sole example of possible employer-generated delay and the earlier comments are probably still valid in respect of other such delays, for example, late information. Indeed, it is thought that the addition of just one example of an employer's delay gives the argument greater force. In the majority of minor Works' contracts the amount of liquidated damages at risk is probably too small to be worth the expense of testing the point.

The plain fact is that a contractor must make a separate application for loss and/or expense. An extension of time has no financial implication other than release from liquidated damages. A contractor may make application for reimbursement of direct loss and/or expense with or without an extension being granted. The courts have emphasised that a claim for loss and/or expense does not depend on a prior grant of extension of time [6]. He may make such financial claim even if he completes the work before the contract date for completion provided only that he can satisfy the architect that he has suffered loss and/or expense due to circumstances which may or may not also be a ground for extension of time.

It can readily be appreciated that this approach is to the contractor's financial advantage in that he is not bound in any way to the grounds for extension of time when claiming financially.

Contractor's duties

It is for a contractor to prove a link between an event and a period of delay. It is not sufficient for the contractor merely to show a delay and point to an event, in effect saying, 'It must have been that' [7].

The contractor has very precise duties under JCT 98 clause 25.2. He must notify the architect in writing as soon as he thinks he is being delayed or that he is likely to be delayed. He must do this whether the delay is the contractor's own fault or due to some other reason. The purpose of this provision is clearly to enable the architect to take whatever steps he may see fit at the earliest possible moment to minimise the effect of the delay. If the contractor fails to so notify the architect, the contractor is in breach of contract [8]. Failure to give notice, however, may not relieve the architect of the obligation to grant an extension in appropriate circumstances. There has been some discussion as to whether the minutes of a site meeting constitute an effective notice. On balance, it is considered that a specific written notice is required [9].

The notice must include the reasons for the delay and state which of the causes is one of the relevant events listed in clause 25.4. As soon as possible, at the same time as the notice if he can, the contractor must send the architect an estimate of the length of the delay beyond the date for completion. The effect of each relevant event must be noted separately, stating whether the delays will operate concurrently. If the contractor thinks the completion date will not be exceeded, he must say so at this point.

If the contractor has made any reference to a nominated sub-contractor, he must send a copy of the written notice, estimates and other details to the sub-contractor. The contractor must send further notices to update his original notice as necessary and copies must be sent to the nominated sub-contractor.

The contractor has two more duties. He must constantly use his best endeavours to prevent delay and he must do everything reasonably required by the architect to proceed with the Works. It has recently been defined as doing 'everything prudent and reasonable' to achieve an objective [10]. This is not, as sometimes thought, an acceleration clause. Neither does it give the architect power to order the contractor to incur additional cost to maintain progress in spite of delays. If it did, the extension of time clause would be redundant. It simply means that the contractor must take all reasonable steps to reduce the effect of a delay; for example, by redeploying part of the workforce to other sections of the work. In practice, it may mean little more than proceeding to work regularly and diligently as far as practicable. Acceleration of the Works can only be achieved by agreement between contractor and employer. If the architect makes a reasonable suggestion, the contractor should take notice, but he is not required to incur any additional expense.

It must be stressed that failure by the contractor to carry out his duties meticulously will not remove the architect's obligation to grant an extension in appropriate instances, but the architect is entitled to take the contractor's failure into account in deciding on the appropriate extension in any particular case. Probably the yardstick is whether and to what extent the architect or the employer could have taken action to reduce the effect upon the completion date if the contractor had given notice at the correct time. The principle is that a person is not entitled to gain as a result of his own wrong [11].

The provisions in IFC 98 clause 2.3 are in broadly similar terms although not so detailed. The obligation to give notice of any delay, to state its cause and to use best endeavours is repeated, but the contractor is not obliged to make an estimate of the extent by which the completion date will be exceeded. Obviously, a prudent contractor will do so and he will supply a very full and well-documented case.

The provisions of MW 98 clause 2.2 are very short. The contractor's obligation is to notify the architect in writing if it becomes apparent that the Works will not be completed by the date for completion for reasons beyond the contractor's control. That is the end of his duties in respect of delays, but if he is wise the contractor will present a full and detailed statement of the causes and an estimate to the architect.

Architect's duties

The architect's duties under MW 98 are equally briefly stated. When he receives the contractor's notice, he must make, in writing, a reasonable extension of time. There is no time limit set, but the architect should act promptly. In general, he should certainly make his decision before the contract completion date is reached, but there may be instances when the delay is continuing. In such a case, the architect may wait until the delay has ended before making an extension.

JCT 98 gives the architect a time limit of 12 weeks from the receipt of all necessary particulars from the contractor in which he must grant an extension of time. If there is less than 12 weeks to the contract completion date, he must make his decision and notify the contractor before the completion date even if the decision is that it is not fair and reasonable to make any extension. In granting an extension, the architect is entitled to take into account any instruction which requires an omission of work and he must so notify the contractor. He must also state which of the relevant events has been taken into account. He must consider two points: whether the delay is caused by a relevant event and, if so, whether it is likely to delay completion.

IFC 98 places no time limit on the exercise of the architect's duty. He is to grant a fair and reasonable extension 'as soon as he is able to estimate the length of delay beyond' the completion date. The architect must, however, make his decision within a reasonable time having regard to all the circumstances. This contract specifically allows the architect to grant an extension of time for certain events in clause 2.4 which occur after the date for completion and which are the fault of the employer. The absence of this provision would lead to such events occurring after completion, rendering time 'at large'. JCT 98 does not contain this provision and it is by no means certain that the architect would have the same power if it were not for the provision noted in the next paragraph. The kind of event referred to includes such actions as the architect issuing instructions for extra work after the date for completion has passed, but while work is still in progress. Such an event would clearly mean further delay.

A valuable provision in JCT 98 allows the architect to make good any deficiencies in previous extensions. After the date for completion has passed the architect may, but no later than 12 weeks after practical completion he must, either:

- fix a later completion date than previously fixed; or
- fix an earlier completion date after taking into account omissions from the work; or
- confirm the completion date previously fixed.

The architect may not, however, fix a date earlier than the original contract completion date. This provision enables the architect to take into account any relevant event which the contractor has failed to notify and which is an employer default, thus preserving the employer's right to deduct liquidated damages.

IFC 98 contains a provision for the architect to make further extensions of time at any time up to 12 weeks after practical completion. There is no provision for reducing any extension previously granted, but otherwise this term serves the same purpose as the term mentioned above in JCT 98. It is sometimes stated that the 12-week period is not mandatory and the architect may carry out his review after that time [12]. Architects would be prudent, however, to ensure that they carry out their duty in this respect within the stipulated period and not rely upon a legal decision which depended upon the special circumstances in which the employer was trying to obtain an advantage from his architect's default.

Relevant events

The architect may only make an extension of time for grounds listed in the contract. JCT 98 refers to these grounds as 'relevant events' (clause 25.4), IFC 98 refers to them as 'events' (clause 2.4). The wording in both contracts is very similar, but there are some important differences. It may be helpful to consider each event separately, indicating where there is a significant difference. In most instances the wording has been shortened and simplified.

Force majeure This is a wider term than 'act of God' meaning broadly 'circumstances independent of the will of humankind'. Most situations which would be covered by this clause, such as strikes, war, etc. are already dealt with under other clauses.

Exceptionally adverse weather conditions This is somewhat broader in meaning than the old term 'exceptionally inclement weather conditions' used in JCT 63. Adverse weather can be rain, snow or frost. It can also be very hot and dry conditions or even high winds. The key word is exceptionally. Thus, heavy snow in January may not be exceptional. In order for this clause to operate the weather must be exceptional having regard to the time of year or the location of the site. It is usual for the architect to require meteorological reports for perhaps 10 years prior to the occurrence.

The contractor is, of course, expected to exercise a reasonable degree of anticipation at tender stage. Therefore, if the date for possession is stated as 25 November, the contractor must allow for the kind of weather which can be expected at that time of year when planning his site operations. In other words, he must expect, for example, snow in December, frost in February, etc. Note that it is the adverse weather which must be exceptional, not the period of time. An architect must consider the contractor's actual progress, even if the contractor is late through his own fault [13].

Loss or damage caused by any one or more of the insurance risks in clause 22 (JCT 98) or 6.3 (IFC 98) This is quite straightforward. The risks (the contract refers to them as 'specified perils') are listed in the contract. They include fire, lightning, explosion, storm, tempest, etc. If progress is delayed thereby, there should be no problem in obtaining a suitable extension. It should be noted that these risks have not been changed by the January 1987 insurance amendments. It should be noted, however, that the risks are more restricted than the wider All Risks.

Civil commotion, strikes, etc. affecting any of the trades employed on the Works or engaged in preparation, manufacture or transportation of goods for the Works A civil commotion is more serious than a riot [14]. Presumably it matters not whether the strike is official or unofficial, but it should be noted that a work to rule does not fall within this clause. The strike must operate directly. A strike which affects a trade which in turn affects another trade involved in the Works does not give grounds for an extension unless, of course, it is a transportation or manufacturing strike such as is specifically covered. For example, a strike in firm A which manufactures door-closers for the site will qualify, but a strike in firm B which supplies a special component of the door-closers to firm A will not, even though progress may be as badly affected by one as by the other.

Compliance with architect's instructions This is a most important clause because it relates to something within the control of the employer. An extension must be granted if the employer is to preserve his right to deduct liquidated damages. It should be quite clear whether an event is a compliance with an architect's instruction, but it should be noted that an instruction to open up work for inspection under clause 8.3 (JCT 98) or 3.12 (IFC 98) will give grounds for an extension only if the uncovered work is found to be in accordance with the contract.

The contractor's work in accordance with the special clause 3.13 of IFC 98, in which the contractor must propose action to ensure that there are no similar failures, will also provide grounds for extension of time on the same basis, although the work itself is to be carried out at no cost to the employer. An architect's instructions under JCT 98 clause 8.4.4 have similar effect. (This provision is discussed in section 1.3.) Instructions regarding the expenditure of provisional sums is also included except where they relate to defined work under SMM 7 (Standard Method of Measurement 7th edition) when the contractor is deemed to have made due allowance in programming and planning in accordance with rule 10.4. (Note that if full information is not provided in the bills of quantities in accordance with rule 10.3, a correction must be made under clause 2.2.2.2 and the correction will be treated like a variation instruction, i.e. the contractor will be entitled to extension of time and to make application for loss and for expense.)

Failure of the architect to comply with the information release schedule or, if not provided, to provide information at the right time This is another clause dealing with events under the control of the employer. This particular relevant event has been changed following the introduction of the information release schedule (see section 4.2). If it has been provided, a simple failure on the part of the architect is sufficient to trigger the event. If it is not provided, compliance with JCT 98 clause 5.4.2 or IFC 98 clause 1.7.2 is the criterion. That is not quite as straightforward.

Delay on the part of nominated sub-contractors or suppliers which the contractor has taken all practical steps to reduce This relevant event appears only in JCT 98 because there are no nominated sub-contractors or suppliers in IFC 98. The clause has been criticised judicially in the JCT 63 form of contract [15] and it is surprising, therefore, that it appears unaltered in the JCT 98 form. The problem is that delay 'on the part of' is not the same as delay 'caused by'. Thus, the contractor is only entitled to an extension under this event if the nominated sub-contractor fails to complete the sub-contract Works within the sub-contract period, or the nominated supplier fails to deliver in accordance with the contract of sale. It matters not why the sub-contractor delays. It may be due to causes outside his control, or his delay may even be malicious. If he exceeds his sub-contract period, the contractor has grounds for extension.

If, however, the sub-contractor completes the work within the period, but he has to return later to correct defects and this causes delay, or if the sub-contractor entirely abandons his work, there are no grounds for extension under this clause. It should be noted that this clause refers only to sub-contractors or suppliers who are nominated. Ordinary (domestic) sub-contractors and suppliers are not covered. Their delays are entirely a matter for the contractor.

The execution or failure by the employer to execute work not forming part of the contract in accordance with clause 29 (JCT 98) or clause 3.11 (IFC 98), or the supply or failure by the employer to supply goods he has agreed to supply This should be perfectly clear. If the employer is to arrange work or materials and his action or inaction causes delay, the contract has grounds for an extension. This clause also covers work carried out by statutory undertakers not in pursuance of their statutory obligations [16]. (See section 3.4.)

The exercise of any statutory power by the UK Government after the date of tender, which directly affects the Works by restricting essential labour or materials or fuel and energy This clause occurs only in JCT 98. The events would probably be covered by force majeure in any case. The key word is 'directly'. Thus, a government order which sets in train a series of events which culminates in such restrictions would not be covered.

The contractor's inability, for reasons outside his control and not reasonably foreseeable at date of tender, to secure labour or materials essential to carry out the Works This clause is present, with slightly different wording, in both forms of contract. However, under IFC 98 provisions, the Appendix may be completed to state that the clause is not to apply. There is no similar provision in JCT 98. In order to have grounds for an extension, the contractor must be able to show that reasonable inquiry, at the time of tender, would not have revealed the problem, and the problem must be caused by factors outside his control. Mere shortage of labour or materials in itself is insufficient.

The carrying out of work or failure to carry out work in pursuance of its statutory obligations, by a local authority or statutory undertaker in relation to the Works This clause relates to the laying of pipes and cables, connections, etc. which the body has an obligation to carry out. The work must relate to the contract Works in order to qualify. Thus, if the electricity supplier causes delay to a site solely because of its operations in regard to a neighbouring site, the contractor has no grounds for extension.

Failure of the employer to give, at the appropriate time, ingress to or egress from any part or the whole site, including passage over adjoining land in the possession and control of the employer, after notice by the contractor in accordance with contract requirements or as agreed by architect and contractor This is not the same as failure to give possession, although it may amount to much the same so far as the contractor is concerned.

In order for the contractor to have grounds for extension under this clause, he must show that he has given notice (if required), the employer is in possession and controls the land and the employer is in default. Therefore, this clause would not apply if roadworks blocked the entrance to the site, or the employer failed to obtain permission for the contractor to obtain access across land belonging to a third party. Depending upon the precise circumstances, the contractor would probably have other remedies in the latter case.

Deferment of possession This clause occurs in IFC 98 and JCT 98. Under MW 98 failure by the employer to give possession to the contractor on the date stated in the contract is a serious breach for which the contractor may be able to claim substantial damages [17]. JCT 98 and IFC 98 sensibly allow the employer to defer possession for a limited time (clauses 23.1.2 and 2.2 respectively) and equally sensibly allow the architect to grant an extension.

Approximate quantities The contractor is entitled to an extension of time where approximate quantities are included in the bills of quantities and the approximate quantities are not a reasonably accurate forecast of the quantity of the work required.

Change in statutory requirements after base date which necessitates alteration to performance specified work This clause only occurs in JCT 98. By clause 6.1.7, this kind of change gives rise to a variation. Therefore, this relevant event may not be strictly necessary, because variations are dealt with under 25.4.5.1. It is worth remembering that there are other statutory requirements besides the Building Regulations.

Use of threat of terrorism or activities of the authorities in consequence Regrettably, this is self-explanatory.

Compliance or non-compliance with clause 6A.1 Under clause 6A.1 the employer has an obligation to ensure that the planning supervisor carries out his duties under the CDM Regulations 1994. The equivalent clause in IFC 98 is 5.7.1. Note that both compliance and non-compliance are covered and the employer has a duty to guarantee performance. Every architect's instruction may result in an extension of time under this clause.

Suspension by the contractor of performance of his obligations This clause puts into effect the extension of time to which the contractor is entitled under section 112(4) of the Housing grants, Construction and Regeneration Act 1996 after he has exercised his right to suspend for non-payment (see section 7.2).

Under both JCT 98 and IFC 98 fluctuations are frozen at contract completion date or any extended date. However, there is an important proviso that the architect must respond to every written notice of delay by granting an extension or otherwise, and the printed text of the extension of time clause must not be amended (except under IFC 98 provisions allowing for deletion of the labour and materials and deferment clauses). If the architect does not respond or amendments are made, fluctuations continue after contractual completion date unless amendments are also made to the fluctuations clauses.

Whatever the contract may say, the contractor has a better chance of obtaining an extension if he presents a clear, well-documented case. A network analysis is invaluable for this purpose.

Liquidated damages

There is much misunderstanding of the liquidated damage clause. Its purpose is to provide for a sum of money to be paid by the contractor for every week (or day) by which the contractor fails to complete the Works after the contract, or extended date for completion. The sum must be a genuine pre-estimate of the likely loss to the employer. Such damages do not have to be proved. A precondition in this case of JCT 98 and IFC 98 is that the architect has issued his certificate of non-completion in accordance with clause 24.1 (JCT 98) or 2.6 (IFC 98). Under IFC 98 if a further extension of time is given, the certificate of non-completion must be cancelled and a new certificate issued. If the architect fixes a new completion date after issuing a certificate of non-completion, the certificate is thereby cancelled and a new certificate is necessary. It has always been considered that another precondition was that the employer should give written notice of intention to deduct or require payment [18]. It is referred to in JCT 98 clause 24.2.1 and in IFC 98 clause 2.7. It also seems to be implicit in clause 24.2.3 in JCT 98 and clause 2.7 in IFC 98 which state that the employer's written requirement remains effective unless withdrawn by him even if the architect has to issue a new certificate of non-completion. It has been suggested that the employer's written notice of intention to deduct may not be a precondition to deduction of liquidated damages and that all that is necessary is that the contractor must be in no doubt that the employer is exercising contractual power to deduct. This line of reasoning suggests that a cheque sent by the employer, in respect of a financial certificate issued by the architect but reduced to take account of liquidated damages, may itself be a sufficient written requirement [19]. The position has been clarified by a case which re-affirms that the employer's written requirement is a precondition [20]. In the written requirement the employer must state that he may require payment or may deduct liquidated damages. He must do this before the date of the final certificate. In addition, of course, if the employer intends to deduct, he must serve the appropriate notice five days before the final date for payment of a particular certificate (see section 5.1). If, unusually, the employer requires the contractor to pay, he must send a special written notice requiring payment at the rate in the Appendix and, if the contractor fails to pay, the money may be recovered by the employer as a debt. The last date for serving either notice is five days before the final date for payment of the final certificate.

The employer may deduct the damages from money due to the contractor. The employer is entitled to deduct even if his loss is less than the stated sum or if he in fact gains by the delay [21]. The liquidated damages clause is often referred to by the contractor as the 'penalty' clause. This is incorrect. A penalty clause is not enforceable. It matters not whether the sum is referred to as a penalty or as liquidated damages [22]. It is the real nature of the sum which counts. A penalty is a punishment. It would be a penalty, for example, if the stipulated sum were greater than the greatest loss which could conceivably follow from the contractor's breach. Another example of a penalty is where the same amount is deductible as damages upon the happening of dissimilar events. In practice, this most often arises because one sum is stated as damages whether the whole or any part of a readily divisible building is delayed [23]. Where liquidated damages in respect of an estate of houses were expressed as £*x* per dwelling per week and no Sectional Completion Supplement was used, it has been held to be inconsistent and unenforceable [24]. The insertion of liquidated damages into a contract requires the greatest care. The wording '£*x* for every week or part of a week' is capable of being construed as a penalty for the obvious reason that damages incurred during a one-day delay are unlikely to be equal to damages during a full week's delay unless, of course, they represented something like rental charges payable at the beginning of each week. Although it is possible to substitute unliquidated for liquidated damages, it cannot be accomplished simply by inserting 'nil' for the amount of liquidated damages. That would simply mean that the employer has specified 'nil' as the total amount of liquidated damages he would suffer as a result of delay [25]. It would be necessary to delete the whole of the liquidated damages clause (clause 24.2 in JCT 98 and clauses 2.7 and 2.8 in IFC 98). The courts will not take account of hypothetical situations when considering this question. They will take a pragmatic approach to whether the sum is liquidated damages or a penalty [26].

In deciding whether a sum is liquidated damages or a penalty, the court would look at the situation at the time the contract was entered into, not at the time of the breach. Provided that the sum is a genuine attempt to estimate the employer's loss in the future, it will be valid. It is of no consequence that it is virtually impossible to estimate accurately, provided that the employer makes reasonable assumptions.

Contractors sometimes say that if there is a liquidated damages clause there must be a corresponding bonus clause for early completion. There is no legal foundation for such statements. It is entirely up to the employer whether he includes a bonus clause. If he does, it need have no monetary relationship to the liquidated damages clause. Such a bonus can be as large or small as the employer wishes. Of course, once the sum is inserted in the contract, it must be paid if early completion is achieved or it may form part of a claim if the contractor is delayed by the employer or the architect.

6.2 Money claims

Under the standard building contracts, to make a claim implies that a contractor considers that he is due an extra payment quite apart from the ordinary system of valuing work done. Claims are a symptom of inefficiency; whether claims are made on the part of the contractor or the architect and employer depends on the circumstances. Contractors generally dislike making claims; architects always dislike dealing with them.

There are three kinds of claim: contractual, common law (sometimes called extra contractual) and *ex gratia*.

Contractual claims are those claims which are made under the express provisions of the particular contract in use. JCT 98 makes provision mainly in clause 26, but also in clauses 13.5 and 34.3.1. IFC 98 deals with them under clause 4.11. MW 98 makes no provision for contractual claims but it does make limited provision for the architect to take direct loss and/or expense into account when valuing variations (see section 4.2). In order to make a contractual claim, the contractor must comply precisely with the terms of the appropriate clause. More details of this are given below.

Common law claims arise outside the express provisions of the contract. They generally relate to breach of implied or express terms of the contract. Thus, such a claim might arise if the employer hindered the contractor's progress on site. Common law claims can also be made in tort, for breach of a collateral contract for *quantum meruit* or sometimes for breach of statutory duty. Some contractors do not realise that it is possible to make a common law claim as an alternative to a contractual claim on the same facts or, if the contractor has not complied with the provisions of the contract claims clause and providing, of course, that the facts constitute sufficient grounds. For example, architects' instructions form grounds for a claim under the contract machinery, but not at common law, because the issue of architects' instructions is not a breach of contract; quite the reverse: it is something expressly allowed by the contract. Contractors should beware, however, that all claims are submitted and dealt with before the final certificate is issued if JCT 98 or IFC 98 are involved. The final certificate under those forms is now conclusive that all claims of whatever kind based on the matters in clause 26 (or clause 4.12 in IFC 98) have been settled.

Ex gratia claims are sometimes known as hardship claims. They have no legal foundation, but the contractor sometimes considers that he has a moral right to extra payment or he has, perhaps, underpriced and he appeals to the employer to help. Whatever the moral rights may be, the contractor's chances with such a claim will usually depend on the benefit to the employer – if any. If to make an ex gratia payment would enable the contractor to continue with a contract when, otherwise, he would go into liquidation, the employer may occasionally make a payment to save himself the expense of securing another contractor to continue the work at a very much increased cost.

JCT 98 clause 26 and IFC 98 clause 4.11 provisions can be considered together because they are very similar in most respects. There is no connection between financial claims – known in the contract as applications for direct loss and/or expense – and claims for extension of time. The contractor must be able to show that the regular progress of the Works has been or is likely to be materially affected by one or more of the list of matters included in the clause. 'Materially' means substantially. Trivial disruptions are excluded. The grounds included are, briefly:

- deferment of possession;
- late receipt of information;
- opening-up of work found to be in accordance with the contract;
- execution of work or supply of materials by the employer;
- failure of the employer to give ingress to or egress from the Works;
- architect's instructions with regard to postponement, discrepancies, provisional sums, variations (and in the case of IFC 98 only, named sub-contractors) (and in the case of JCT 98 only, provisional sums for performance specified work);
- where approximate quantities are included in the bills of quantities and the approximate quantities are not a reasonably accurate forecast of the quantity of work required;
- compliance or non-compliance with duties in relation to the CDM Regulations;
- suspension of the contractor's performance of obligations.

Contractor's duties

In order to be able to claim, the contractor must have suffered or be likely to suffer direct loss and/or expense. This means damages as generally understood. The word 'direct' is important. Direct damage flows naturally from the breach without any other intervening cause [27]. Therefore, if the architect issues an instruction varying the type of ironmongery and the contractor incurs loss and/or expense, over and above the difference in the value of the work and materials, he is entitled to claim. But if the cause of the contractors loss and/or expense is the fact that the supplier fails to deliver the new ironmongery according to an agreed delivery date, the supplier's failure would be an intervening cause and the contractor could not cite the architect's instruction as giving rise to direct loss and/or expense. Direct damage also means the normal and ordinary damage which does depend upon special circumstances. In order to claim such special damages in addition to the direct damages, the contractor would have to proceed at common law and show that the special circumstances were known to the parties at the date the contract was entered into [28]. The precise circumstances of every case require careful study.

It is vital that the contractor makes his application to the architect as soon as he is aware that the regular progress is being or is likely to be affected. The architect is entitled, indeed he has a duty, to reject late applications. The reason for this provision is clear. The architect must have the opportunity, so far as possible, to issue instructions to overcome the difficulty. He cannot do this unless he is aware of it. Therefore, the contractor must not delay making application simply because he has not marshalled all his facts. It is not considered, however, that an architect would be justified in rejecting an application if the fact that it was technically late did not prejudice the employer's position in any way.

The initial application can be quite simple:

> *Dear Sir,*
>
> *We hereby make application under clause 26.1* (substitute '4.11' when using IFC 98) *of the conditions of contract as follows:*
>
> *We have incurred/are likely to incur direct loss and/or expense and financing charges in the execution of this contract for which we will not be reimbursed by a payment under any other provision in this contract* because the regular progress of the Works has been/is likely to be materially affected by* (describe), *being a matter in clause* (insert one or more of clauses 26.2.1 to 26.2.8 when using JCT 98 or clause 4.12.1 to 4.12.8 when using IFC 98).
>
> *Yours faithfully,*

If deferment of possession is involved, continue as follows from the asterisk: *due to deferment of giving possession of the site.*

Architect's duties

It is for the architect to decide whether the claim is valid. In order to be able to form an opinion, he will usually request further information which the contractor must provide as soon as reasonably possible. If the architect decides that the claim is valid, he must then ascertain the amount of direct loss and/or expense or ask the quantity surveyor to do so. Note that it is not up to the quantity surveyor to decide validity, only the architect can do this [29]. The contractor must provide all the information required by the architect or the quantity surveyor to enable them to carry out the ascertainment.

The contractor must make a new application whenever a fresh matter arises. If, however, the matter referred to in the original application is continuing, the architect's duty is to continue to carry out ascertainment, certifying such amounts as soon as each ascertainment is complete. The architect may not simply wait until the end of the contract before certifying such sums. If the architect fails to act, it is a breach for which the employer is responsible [30].

Key points

The presentation of the supporting information for a claim requires and deserves careful thought. Some contractors appear to think that if they send large parcels of documents, the architect will be obliged to certify something approaching the sum required. However, so long as the architect is baffled by the claim, he will not make a decision. The correct approach is to make a claim as simple and easy to digest as possible. The main points should be logically arranged, based on and referenced to contract clauses. Each point should be cross-referenced to evidence bundled separately.

The contractor who wishes to avoid long delays and numerous requests for further information will submit his supporting information in this way as soon as possible after making the initial application, without waiting for the architect to request it.

What may be included in a claim? The answer is any direct loss and/or expense directly referable to one or more of the matters in the clause. What constitutes direct loss and/or expense depends upon the particular circumstances. It can become quite complicated, but attention should be given to the following:

- Plant and labour inefficiency as a direct result of the disruption.
- Increases in cost occurring during the period of disruption.
- Increases in head office overheads. This is a difficult cost item to disentangle and many contractors put forward costs based on a formula. The most used is the Hudson formula, but Eichlay and Emden formulae are also used. Generally, however, it is better to produce proper records [31].
- Establishment costs calculated at the time of disruption.
- Loss of profit is a permissible part of a claim [32]. The contractor must be able to show that he would have been able to earn the profit elsewhere but for the disruption [33].
- Interest and financing charges are allowable provided they are a direct result of the disruption [34].
- Cost of preparing the claim cannot normally be claimed except in so far as it forms part of head office overheads. There appears, however, to be no good reason why a contractor should not be able to claim, as special damages, the cost of preparing a claim to satisfy an unreasonable architect or quantity surveyor of the justice of a claim [35].

The key point to be borne in mind is that loss and/or expense can be claimed provided it is a foreseeable and direct result of the disruption or delay. Moreover, although the contractor must prove that he has suffered or incurred the loss before he is entitled to reimbursement, the burden of proof is 'the balance of probabilities'. This is very much less than the standard of 'beyond a reasonable doubt' in criminal cases. The balance of probabilities is like saying that it is more likely than not that the contractor suffered the loss.

If there is an extremely complicated interaction between a number of matters included in the claims clause such that it is virtually impossible to separate the costs into a series of items, it is possible to present the evaluation on a global basis [36]. In such cases it is important to ensure that there is no duplication and to identify individual causes where this is possible.

The distinction must be made between global evaluation, which may be permitted in appropriate circumstances, and the global approach to liability, which is not permitted [37]. A contractor must frame his application with sufficient detail that the architect is able to see exactly to what matters the contractor is referring and in what way each of the matters delayed or disrupted the Works. For example, it is not good enough for the contractor simply to say: 'There was a multitude of variations throughout the project, AIs (architect's instructions) were late, the employer's men were constantly tramping through the site and the result was an overrun of six weeks and £24,000 extra loss and expense.' Instead, he should list each variation, when it was received and demonstrate the effect of each one individually, and he should do the same with the late AIs and the occasions when the employer's men disrupted the Works. If the claim is for costs due to delay and the consequences of numerous events have a complex interaction, it may be permissible to maintain a composite claim if it is impossible to identify the specific relation of each event and its time and cost result [38]. The circumstances when this approach will be acceptable will be rare in practice. A contractor is entitled to submit a global claim if he wishes, but he should be aware that there are serious difficulties of proof [39]. Global claims, if presented as 'all or nothing', will fail completely if some of the causative events are not established. Therefore, this type of claim should show how the total amount is to be reduced if some events are not established and to take account of the contractor's own inefficiencies and issues of that nature [40].

It often happens that a contractor suffers loss and/or expense due to a variety of causes, some of which are permissible as contractual claims, others of which fall outside the claims clause and can only form the basis of a claim at common law. Strictly common law claims must be taken through the courts, but it usually does no harm to submit them to the architect as part of an overall claim provided it is made clear that they are not contractual claims. The architect should refer all common law claims to the employer who might well decide that it is more convenient to settle them through the architect than through arbitration or the courts. Of course, any money found to be due to the contractor as a result of a common law claim cannot be certified through the contract. It must be paid by the employer separately. All claims for loss and/or expense under MW 98, with the very limited exception already noted, are common law claims. Finally, it should be noted that whether the contractor is claiming through the contract or at common law, he can only claim – and the architect can only ascertain – the precise loss and expense suffered [41].

References

1. Dodd v Churton (1897) 1 QB 562.
2. Wells v Army & Navy Co-operative Society Ltd (1902) 86 LT 764.
3. Percy Bilton Ltd v Greater London Council (1981) 20 BLR 1.
4. Peak Construction (Liverpool) Ltd v McKinney Foundations Ltd (1970) 1 BLR 111.
5. Scott Lithgow v Secretary of State for Defence (1989) 45 BLR 1.
6. H Fairweather & Co Ltd v London Borough of Wandsworth (1987) 39 BLR 106.
7. Ascon v Alfred McAlpine (1999) unreported.
8. London Borough of Merton v Stanley Hugh Leach Ltd (1985) 32 BLR 51.
9. John H Haley Ltd v Dumfries & Galloway Regional Council (1988) GWD 39–1599.
10. Victor Stanley Hawkins v Pender Bros Pty Ltd (1994) 10 BCL III.
11. Alghussein Establishment v Eton College [1988] 1 WLR 587.
12. Temloc Ltd v Errill Properties Ltd (1987) 39 BLR 30.
13. Walter Lawrence & Son Ltd v Commercial Union Properties (UK) Ltd (1984) 4 Con LR 37.
14. Levy v Assicurazioni Generali [1940] AC 791.
15. J Jarvis & Sons Ltd v Westminster Corporation (1970) 7 BLR 64.
16. Henry Boot Construction Ltd v Central Lancashire New Town Development Corporation (1980) 15 BLR 1.
17. The Rapid Building Group Ltd v Ealing Family Housing Association Ltd (1984) 1 Con LR 1.
18. A Bell & Son (Paddington) Ltd v CBF Residential Care & Housing Association (1989) 5 Const LJ 194.
19. Jarvis Brent Ltd v Rowlinson Constructions Ltd (1990) 6 Const LJ 292.
20. Finnegan v Community Housing (1993) 65 BLR 103.
21. Clydebank Engineering v Yzquierdo Castenada [1905] AC 6.
22. Dunlop Pneumatic Tyre Co Ltd v New Garage Motor Co Ltd [1915] AC 79.
23. Stanor Electric Ltd v R Mansell Ltd (1987) CILL 399.
24. Bramall & Ogden Ltd v Sheffield City Council (1983) 1 Con LR 30.
25. Temloc Ltd v Errill Properties Ltd (1987) 39 BLR 30.
26. Phillips Hong Kong Ltd v The Attorney General of Hong Kong (1993) 9 Const LJ 202.
27. Croudace Construction Ltd v Cawoods Concrete Products Ltd (1978) 8 BLR 20.
28. Hadley v Baxendale (1954) 9 LR Ex 341.
29. John Laing Construction Ltd v County & District Properties Ltd (1982) QB 24 November.
30. Croudace Ltd v London Borough of Lambeth (1986) 6 Con LR 70.
31. Tate & Lyle Food and Distribution Ltd v Greater London Council [1982] 1 WLR 149.
32. Saint Line Ltd v Richardson, Westgarth & Co Ltd [1940] 2 KB 99.
33. Peak Construction (Liverpool) Ltd v McKinney Foundations Ltd (1970) 1 BLR 111; AMEC Building Contracts Ltd v Cadmus Investments Co Ltd (1997) 13 Const LJ 50; City Axis v Daniel P Jackson (1998) CILL 1382.
34. Rees & Kirby Ltd v Swansea City Council (1985) 5 Con LR 34.
35. Babcock Energy Ltd v Lodge Sturtevant Ltd (formerly Peabody Sturtevant Ltd) (1994) unreported.
36. J Crosby & Sons Ltd v Portland UDC (1967) 20 BLR 34.
37. Wharfe Properties Ltd v Eric Cumine Associates (1991) 52 BLR 1.
38. Mid Glamorgan County Council v J Devonald Williams and Partner (1991) 8 Const LJ 61.

39. GMTC Tools and Equipment Ltd v Yuasa Warwick Machinery Ltd (1994) 73 BLR 102.
40. How Engineering Services Ltd v Lindner Ceilings Partitions plc. 17 May 1995 unreported.
41. Alfred McAlpines Homes North Ltd v Property and Land Contractors Ltd (1995) 76 BLR 65.

7 The end

7.1 Practical completion and defects liability

When the architect is of the opinion that practical completion has been achieved and the contractor has provided all the information necessary for the health and safety file, he must issue a certificate to that effect, stating the date of practical completion (JCT 98 clause 17, IFC 98 clause 2.9 and MW 98 clause 2.4).

The contractor is always anxious to obtain this certificate which confers enormous benefits. It marks the date when:

- The defects liability period begins.
- The contractor's liability for insurance ends.
- The contractor's liability for frost damage ends.
- The employer's right to deduct full retention ends and half the retention already held is due for release.
- Liability for liquidated damages ends.
- The contractor's licence to be on the site ends.
- Regular interim certificates cease.
- The period of the review of extensions of time begins (not applicable to MW 98).

Such is the importance of practical completion that the contractor might be forgiven for thinking that the term would be carefully defined in the contract. Unfortunately, no definition is to be found. This does not mean that the architect is completely free to fix any date he chooses. The date is left for his opinion, but that opinion is open to question if the contractor chooses to refer the point to arbitration. A number of cases have considered the matter, not always coming to the same conclusion. However, a number of important factors can be identified. Practical completion does not mean substantially or nearly complete. It means complete except for minor items when there are no apparent (i.e. visible) defects [1]. One judge held it to mean complete for the purpose of allowing the employer to take possession and use the Works as intended. That may be too relaxed, but he considered that if it meant completion down to the last detail it would be a penalty clause and unenforceable [2]. A working definition might be that practical completion is when there are no defects apparent and when such minor items as are left to be completed can be completed without any inconvenience to the employer using the building as intended. A contractor who feels that the practical completion certificate is being unreasonably withheld should inform the architect in writing, noting the date on which the contractor considers practical completion has been achieved.

On many projects, it is common to have what is known as a 'handover meeting', at which the architect, the employer and consultants meet the contractor and jointly inspect the building. In some instances, the decision that the building has reached practical completion is effectively left to the employer. Such an arrangement is quite wrong and if the employer were to make the final decision, it would be a breach of contract which stipulates that the decision is one for the architect alone. If the employer interferes with or obstructs the issue of any certificate, the contractor has grounds for determination (JCT 98 clause 28.2.1.2, IFC 98 clause 7.9.1(b) and MW 98 clause 7.3.1.2).

Whether there would be much point in determination at such a late stage in the contract is a matter to be decided on the particular circumstances. It is probably more usual that the employer is anxious to accept the building, often against the architect's advice. Since practical completion is greatly to the contractor's advantage, interference by the employer to secure the early issue of the certificate would seldom be questioned by the contractor.

Sometimes, the employer will take possession of the building some time before the architect would have certified practical completion. The contractor does not dissent and both employer and contractor expect the architect to issue a certificate of practical completion. The architect may consider that, if both parties agree, he is free to issue the certificate. This is not, in fact, the case: the contract requires the architect to issue the certificate only when certain criteria have been satisfied. The certificate is the formal expression of the architect's opinion [3]. If the architect issued a certificate which did not represent his opinion, he would certainly be guilty of unprofessional conduct. Quite apart from matters of professional ethics, there could be serious legal consequences for the architect depending on particular circumstances, despite the agreement of the parties to the contract.

In some instances, the decision seems to be taken by the clerk of works. Although this appears to be wrong, it must be remembered that, in forming his opinion, the architect is entitled to take into account any information he chooses, including the views of the clerk of works. The opinion, however, must be that of the architect. A practical completion certificate signed by the clerk of works is simply a useless piece of paper.

Much more serious are the situations where the contractor suspects that a certificate of practical completion is being delayed, because the employer/developer has been unsuccessful in leasing the property and he opts for the liquidated damages until a lease can be arranged. The contractor should be able to rely upon the architect to act properly in accordance with his duty under the contract even though considerable pressure may be imposed by the employer.

Defects liability

Once practical completion has been certified, the defects liability period begins. The length of the period can be anything, provided it is clearly stated in the contract. A common period is six months, but many architects consider that 12 months is more appropriate and mechanical installations commonly specify 12 months on all contracts. It should be noted that none of the three contracts under consideration allows for differing periods to be inserted for different parts of the same building. Problems can also occur if the sub-contract and main contract periods do not correspond.

The defects liability period is often mistakenly referred to as the 'maintenance period'. Maintenance implies that the building must be kept in pristine condition: floors polished, paintwork renewed, etc. Contractors sometimes think that they have a duty to attend to any matter raised by the architect during this period. That is not the case. There is a slight difference in wording between the three contracts, but in essence the contractor must make good any defects, shrinkages and other faults which appear during the period (JCT 98 clause 17.2 and IFC 98 clause 2.10). MW 98 clause 2.5 refers to 'defects, excessive shrinkages or other faults', which is slightly less onerous on the contractor because minor shrinkages would appear to be excluded.

The phrase 'other faults' is to be interpreted to mean faults which are like defects and shrinkages. 'Defects' means work not in accordance with the contract and not defects due to some other reason, such as poor standard of specification or the employer's occupancy. 'Shrinkages' are only to be made good if they are due to materials or workmanship not being in accordance with the contract. Thus, if the moisture content of the timber was specified too high, but the timber was fixed at the specified moisture content and shrinkage took place solely because, say, central heating reduced the moisture content below the specified level, the contractor would not be liable.

The contractor is also liable for frost damage, provided the frost occurred before the date of practical completion. Damage due to frost occurring after practical completion is the employer's liability.

Reference is made to defects which 'appear' during the period. The word seems to extend to mean defects which are apparent during the period [4]. In any event, the contractor will be liable for any disconformity between the Works and the contract documents.

The architect has no right to instruct how the defects are to be made good. If he does, the contractor may be entitled to payment as though the architect had issued a variation [5].

MW 98 is not specific, but JCT 98 and IFC 98 empower the architect to issue instructions that defects are to be made good at any time during the period and up to 14 days thereafter. The contractor is not entitled to wait until the end of the period before he commences the making-good process. JCT 98 requires the architect to specify the defects in a schedule which must be delivered within 14 days after the end of the defects liability period. He is not entitled to issue any further instructions under the defects liability clause after the issue of the schedule or after the end of the 14 days, whichever is earlier.

The import of this is sometimes misunderstood by contractors. The contractor can refuse to make good defects after the end of the period, but if work or materials are not in accordance with the contract, they are still the contractor's liability. The defects liability period is largely for the contractor's benefit. If it were not for this period, the contractor would have no right to return to the site and rectify defects. The employer could simply take arbitration proceedings for damages (probably the cost of having the work put right by others). This is the position after the end of the period. In general, the architect will refer any defects which appear after the defects liability period to the contractor, in the hope of obtaining speedy rectification. If the contractor refuses, and the defects are a result of the work or materials not being in accordance with the contract, the employer can employ others and arbitrate against the contractor. The employer has the right to charge the contractor the full cost of putting right defects which appear after the end of the defects liability period.

The contractor is to make good the defects at his own cost. The architect is empowered, however, to instruct that some or all of the defects are not to be made good (with the consent of the employer) in which case an appropriate adjustment may be made to the contract sum. The precise meaning of 'appropriate deduction' has been the subject of some debate. It is considered that it refers to something less than the cost of having the work done by others, otherwise it would have been described in the same way as work done by others if the contractor does not comply with an instruction (JCT 98 clause 4.1.2), i.e. 'all costs incurred in connection with ...' Such an instruction may be issued if the employer does not want to be disturbed by the rectification of minor faults or because the architect considers that it would be prudent to employ someone else to do the work. The defects liability clause gives the contractor the right to reduce the cost of remedial works by doing them himself. It is now established that where the employer elects not to have the defects corrected by the contractor, the sum he can recover (appropriate adjustment) is limited to the sum which represents the cost which the contractor would have incurred if he had been called upon to remedy the defects [6]. It now appears that if defects which appear during the defects liability period are not notified to the contractor until later, it is still a breach of contract on the part of the contractor, but the architect cannot insist on the contractor putting the work right and the employer can only recover from the contractor what it would have cost the contractor to correct the defects [7]. If some of the defects are the responsibility of the employer, the architect may instruct that such defects are to be made good, but that the contractor is to be paid for so doing. It is not thought that the contractor is obliged to deal with such defects even if paid. At the end of the defects liability period when all defects have been made good, the architect must issue a certificate to that effect. Under JCT 98 terms the second half of the retention will be released after the issue of this certificate.

Partial possession

MW 98 contains no provision for the employer to take partial possession of the Works. JCT 98 clause 18 and IFC 98 clause 2.11 refer to partial possession. It is important to remember that partial possession is not the same as phased or sectional completion. If the employer wishes the Works to be completed according to a particular timetable, he should incorporate the sectional completion supplement which is designed for this purpose. If the standard form is used without any special amendments, the employer cannot achieve a phased completion by simply – as is sometimes done – setting out the phases and completion dates in the bills of quantities. The contractor's obligation in such a case is simply to complete by the date for completion in the Appendix to the printed form. Any conflicting dates in the bills are overridden by the provisions of JCT 98 clause 2.2.1 (and IFC 98 clause 1.3).

The purpose of the partial possession clause is to enable the employer, with the consent of the contractor, to take possession of any part or parts of the Works before practical completion of the whole of the Works. Note that the employer cannot take possession without the contractor's consent (the contractor is in possession of the site), but the contractor may not refuse such consent unreasonably. If the contractor agrees to the employer taking possession, he should take care that the procedure is carried out formally. He should give consent in writing subject to the architect issuing a written statement stating the date on which partial possession was taken in accordance with clause 18.1.1 (IFC 98 clause 2.11). The architect must issue the statement as soon as partial possession is taken, but the contractor would be advised to secure the statement no later than the date of possession and make the handing-over of the written statement a condition for the giving of consent. The reason for taking such great care is that it has been held that if the employer merely enters the Works, stores materials and even carries out factory processes in a building before practical completion, but without the issue of any formal certificate or written statement, he has not taken partial possession. Therefore, if there is a fire which destroys the building and the contents brought in by the employer, then the contractor is liable for the cost unless the position has been agreed with the insurers [8].

The issue of the architect's written statement, however, will ensure that the appropriate contractual provisions come into operation as follows:

- The defects liability period is deemed to have begun, for the part taken into possession, on the date in the written statement.
- When all defects, etc. have been made good, a making good of defects certificate for the part is to be issued.
- The contractor's obligation to insure the part ceases (this is a crucial provision because the responsibility becomes the employer's).
- The amount of liquidated damages which may be payable is reduced in proportion to the value of the part.

Provided that the contractor is not inconvenienced by the partial possession of the Works and he secures the architect's written statement, there are advantages in acceding to the employer's request.

JCT 98 clause 23 and IFC 98 clause 2.1 contain provisions to cover the situation where the employer wishes to make use of or occupy the Works before practical completion for storage or other purposes, but does not wish to take partial possession. The contractor's consent is required, but he is not to withhold such consent unreasonably. Before the contractor gives his consent the party responsible for Works insurance must obtain from the insurers confirmation that the proposed use will not prejudice the insurance. If the contractor is responsible for insurance and the insurers require an additional premium, the contractor must notify the employer, and if the employer still requires the use of the Works, the amount of the additional premium payable by the contractor must be added to the contract sum.

7.2 Suspension and determination

In response to the Housing Grants, Construction and Regeneration Act 1996, clause 30.1.4 has been inserted into JCT 98 (IFC 98 clause 4.4A and MW 98 clause 4.8 have the same effect). The clause is said to be without prejudice to any other rights and remedies which the contractor may possess. This means that if the contractor exercises this right, it does not prevent him from exercising other rights under the contract or at common law.

If the employer fails to pay the contractor in full the amount due by the final date for payment, the contractor may issue a notice to the employer stating his intention to suspend. It must also state the grounds of the suspension (i.e. that the employer has failed to pay a particular sum of money which should have been paid by a specified date). The notice must be written and it must allow the employer seven days in which to pay the money properly due. Obviously, the employer is not obliged to pay any amount for which he has issued a valid notice saying that he intends to withhold payment. It should be noted that in order to comply with the clause, the contractor must send a copy of the notice to the architect. This is not a requirement of the Act and, if the contractor omitted to send a copy to the architect, it would not stop him from being entitled to suspend work.

This is a draconian remedy, but no doubt merited on occasion. It is suggested that the notice be sent by special or recorded delivery to ensure that it is received and that there is evidence to that effect. The suspension may continue until payment has been made in full. The clause goes on to say, reflecting the relevant section of the Act, that the suspension should not be treated as a suspension to which the determination clause refers, neither is it to be treated as a failure to proceed regularly and diligently. That should go without saying.

All JCT contracts provide the contractor with a further remedy to deal with the period of suspension. It is to be taken into account for an extension of time and the contractor is also entitled to make applications for loss and/or expense.

Under the general law it is possible to discharge a JCT contract in one of four ways:

- By performance, when both parties have completed their obligations under the contract.
- By agreement, when both parties agree that, despite not having completed or perhaps started their obligations to each other, the contract should come to an end. This should be done with the same formality with which the original contract was entered into. Unless executed as a deed, such an agreement will usually be effective only if both parties have outstanding obligations.
- By frustration, when events outside the control of both parties render the contract something radically different from what was originally contemplated [9].
- By breach, when one of the parties commits a breach so serious as to indicate a clear intention not to be bound by the contract and the other party accepts that the contract is at an end. There are serious dangers for the party accepting the repudiation, which will be discussed later.

All three contracts state that the contractual right to determine is without prejudice to any other rights and remedies which the employer or contractor may possess. In other words, the employer or the contractor may opt to bring the contract to an end, if they have sufficient grounds, using common law principles rather than the contractual machinery. In certain instances it may pay them to do so because determination at common law entitles the party properly carrying out the determination to damages; whereas determination under the contract simply entitles him to whatever remedies are stipulated in the contract [10]. However, it should be noted that some of the grounds listed in the contract entitling the employer or contractor to determine the contractor's employment would not be sufficient to justify determination of the contract at common law.

The three contracts refer to determination of employment, thus the contract itself is still in existence and provides for the arrangements to be made after determination. If determination of the contract takes place at common law, the whole contract is extinguished and none of its clauses can apply thereafter except the dispute resolution clauses.

The determination clauses in JCT 98 are: 22C, 27, 28 and 28A. In IFC 98 they are clauses 6.3C.4.3 and 7. In MW 98 determination is covered by clauses 5.5 and 7.

Damage by insured risks

JCT 98 clause 22C and IFC 98 clause 6.3C.4.3 may be considered together because they are identical. They deal with the situation where loss or damage is caused to the Works by one or more of the insured risks if the insurance is the responsibility of the employer and the Works are alterations or extensions to an existing building.

The contractor must give written notice to the employer as soon as he discovers the loss or damage, irrespective of whether he intends to implement the determination provision. The contract states that 'if it is just and equitable so to do', either party may serve notice on the other to determine the contractor's employment. The notice must be served by special or recorded delivery within 28 days of the occurrence and then the other party has seven days within which to refer to the dispute resolution procedures as to whether the determination is 'just and equitable'. If notice to refer is not given within the seven days, the right is lost.

Three points deserve expansion:

- The meaning of 'just and equitable' is not clear. The provision is presumably intended to cover the position where alterations are being carried out to an existing building which is itself badly damaged by the occurrence. In such circumstances it would seem pointless to proceed with a contract for alterations if the building to be altered is largely destroyed. Such an event would probably rank as frustration under the general law in any case.
- Where a notice is to be served by special or recorded delivery, it is

wise to follow the direction precisely. It is probable that a properly receipted personal service would be valid [11], but it is not entirely certain. When either party takes a serious step under the contract, it is safest to comply with the provisions exactly to the extent of repeating the wording in the contract in any notice required to be given, so that there can be no doubt about intentions. The courts tend to take a business-like view of such items as notices, and may overlook minor imperfections, looking at the spirit rather than the letter, unless it is thought that one of the parties is seeking unfair advantage, when a notice may be interpreted very strictly.

- The determination notice must be given within 28 days of the occurrence, not discovery of the occurrence. Although it seems unlikely that damage of sufficient seriousness to warrant determination under this clause will lie undiscovered for more than a day or so, it is quite conceivable that by the time the contractor has found the damage and notified the employer, the employer may have only three weeks in which to assess the problem and decide what to do.

The consequences of determination under this provision are the same as if the contractor determines his own employment under JCT 98 clause 28A or IFC 98 clause 7.9 except that the contractor is not entitled to receive any payment in respect of direct loss or damage arising from the determination.

Determination by employer

JCT 98 clause 27 and IFC 98 clauses 7.2–7.5 deal with determination by the employer. They are similar and can be considered together. The equivalent provision in MW 98 (clause 7.2) will be discussed separately.

The grounds for determination are if the contractor:

- wholly or substantially suspends the carrying out of the work before completion without reasonable cause;
- fails to proceed regularly and diligently;
- refuses or neglects to remove defective work, etc. after notice from the architect and the Works are materially affected;
- fails to comply with clauses restricting sub-letting, restricting assignment or dealing with named persons as sub-contractors (IFC 98 only);
- fails to comply with the CDM Regulations;
- becomes insolvent or makes an arrangement with creditors, etc.;
- is guilty of corruption, i.e. giving or receiving bribes and the like.

In practice, it may be difficult to show that the contractor has wholly suspended without reasonable cause or failed to work regularly and diligently. The employer seeking to rely upon these grounds must be very sure of his facts. It has been suggested that a term in a contract requiring a contractor to proceed with due diligence is an obligation to execute the Works so that key dates and the completion date will be met [12]. Since most contracts, in the absence of sectional completion provision, will not have key dates other than the completion date, it seems on that basis that the contractor's contention that he is working regularly and diligently will be difficult to disprove unless it is clear that he cannot meet the completion date. Architects who fail to take action in respect of a contractor who fails to proceed regularly and diligently may risk legal action by the employer. 'Regularly and diligently' has been considered by the courts and it has been stated that it means 'essentially to proceed continuously, industriously and efficiently with appropriate physical resources so as to progress the Works steadily towards completion substantially in accordance with the contractual requirements as to time, sequence and quality of work' [13] . Certainly, it is not enough if the contractor merely fails to work according to programme or stops working on part of Works for a brief period.

It is difficult to understand why the third ground is included at all. The employer already has a satisfactory remedy if the contractor fails to comply with the architect's notice requiring compliance with an instruction (see section 4.2). In order for this ground to be effective, the contractor must refuse or neglect to deal with defective work. That is, he must fail to act despite reminders from the architect. Moreover, his inaction must result in the Works being materially affected – presumably if the defective work would have an adverse effect upon what followed. It is not thought that failure to remedy defects which are not urgent would come within this ground. The Amendments to JCT 80 and IFC 84 in 1987 and 1988 respectively removed the requirement that the neglect must be persistent. The argument was apparently that the proviso that determination should not be unreasonable or vexatious would prevent the employer determining if the contractor neglected to comply with an instruction on just one occasion. This argument appears to be flawed, because if compliance with the instruction was sufficiently important, it is quite possible for determination to be reasonable after one or two instances of neglecting to comply on the part of the contractor. This is a point to be particularly aware of.

The fourth ground is intended to give the employer a remedy if the contractor does not comply with the clauses noted. It is thought that the contractor's failure would have to be quite unequivocal. In most cases the failure could be dealt with less drastically by an appropriately worded letter.

The fifth ground was introduced to give the employer a remedy if the contractor fails to carry out his contractual duties in regard to the CDM Regulations. Presumably, a grave lapse is envisaged before the provision would be applied.

The procedure in regard to determination following the contractor's insolvency and some other matters have been changed. It is convenient to deal with them together.

Determination is still automatic, but may be reinstated if the employer and the contractor agree, where a provisional liquidator or a trustee in bankruptcy has been appointed, or where a winding-up order has been made, or if the contractor has passed a resolution for a voluntary winding-up (except for reconstruction). However, in other instances – for example, where the contractor has made a composition or arrangement with creditors – determination is not automatic, but the employer has the right to serve a notice of determination at any time after the event. From the date on which he could have given notice, the employer is not bound to make any further payments and the contractor is not bound to carry out any more work. Clauses 27.5 and 7.5 of JCT 98 and IFC 98 respectively contain complex provisions which allow the employer and the contractor to make an interim or a permanent arrangement for the work to be continued, presumably on terms acceptable to both parties. Clauses 27.5.3 and 7.5.3 stipulate that during the period while an interim arrangement is in place, the employer may not make any set-off against the contractor. An important power for the employer permits him to take measures for adequate protection of the Works and materials. The contractor may not hinder these measures and the employer is entitled to deduct the cost from any money becoming due to the contractor. The employer may determine on grounds of corruption by simply serving a notice of determination stating the ground.

In order to determine on any of the first four grounds, the architect must first serve a notice by special or recorded delivery specifying the default. Notice may be served also by actual delivery (note that service by fax is not included). The contractor has 14 days from receipt of the notice in which to remedy or start to remedy the default. If he takes no action, the employer may then serve notice of determination by actual, special or recorded delivery. He has 10 days in which to act. Note that the architect is not empowered to serve the actual notice of determination. There is a sting in the tail for the contractor because, if he remedies the default after receiving the default notice and some time later repeats the default, the employer is entitled to serve notice of determination without first serving a further default notice. To what extent the repeated default must be exactly the same as the original default is a matter of speculation, but it is wise to avoid being a test case.

There is an important proviso to the effect that notice of determination must not be given unreasonably or vexatiously. Thus, it would be unreasonable to attempt to take unfair advantage [14] and vexatious to serve notice in an attempt to annoy.

MW 98 allows the employer to 'cancel this contract' if the contractor is guilty of corruption (clause 5.5). Clause 7.2.1 entitles the employer to serve notice of determination on four of the grounds already mentioned, i.e. wholly suspending the Works, failure to work diligently, failure to comply with the CDM Regulations and insolvency of the contractor. The earlier comments on these grounds apply equally to MW 98. A default notice is specified to be given by the architect in respect of the first two grounds. The contractor has seven days from receipt to end the default, failing which the employer may determine by giving a further notice. In the case of insolvency, no default notice is required and notice of determination is effective on the date of receipt by the contractor. In view of the employer's difficulties in determining on these grounds, mentioned earlier, a contractor might be forgiven for thinking that, short of corruption or insolvency, he is safe from determination provided that he stays on site and does at least some work. Such a view may be unduly optimistic in the light of current law. The various grounds for determination are briefly summarised in Table 7.1.

Determination by contractor

Failure to pay is a common ground to all three contracts under discussion. This is a valuable option for the contractor. At common law failure to pay is not grounds for repudiation unless the circumstances are exceptional [15]: the remedy is to use the dispute resolution procedures.

Before issuing the determination notice, the contractor must send a notice specifying the default to the employer. It is a 14-day notice under JCT 98 (clause 28.2.3) and under IFC 98 (clause 7.9.3) and a seven-day notice under MW 98 (clause 7.3.1). All notices must be given by actual, special or recorded delivery and the remarks on delivery in relation to employer determination are applicable.

Under JCT 98 and IFC 98 the employer's obstruction of the issue of any certificate (not just financial certificates) is a ground for determination. Under MW 98 the obstruction may additionally relate to the carrying out of the work or failure to make the site available on time. It is usually difficult to prove that the employer has interfered with the issue of a certificate unless the architect is ill-advised enough to say so in a letter refusing issue. It would be sufficient if the contractor could show that the employer had instructed the architect to reduce the amount on a financial certificate or requested him to delay the issue of, say, a certificate of practical completion [16]. Failure by the employer to comply with the contractual provisions regarding the CDM Regulations is a common ground to all three contracts.

If the Works are suspended for a period of one month (in the case of IFC 98 and MW 98) or for the period inserted in the Appendix (in the case of JCT 98) for any of the reasons listed, the contractor can determine. Most of the reasons can be readily identified from Table 7.1 above and read in detail in the particular contract form. The reason is irrelevant

Table 7.1 Grounds for determination under JCT 98, IFC 98 and MW 98

Ground	*JCT 98*	*IFC 98*	*MW 98*
By employer			
Contractor wholly or substantially suspends work	X	X	X
Contractor fails to proceed regularly and diligently	X	X	X
Contractor neglects defective work	X	X	
Contractor assigns without consent	X	X	
Contractor sub-lets without consent	X	X	
Contractor fails to comply with CDM	X	X	X
Contractor fails to comply with named person clause		X	
Contractor becomes insolvent, etc.	X	X	X
Contractor guilty of corruption	X	X	X
By contractor			
Employer fails to pay including VAT	X	X	X
Employer obstructs certificate	X	X	X
Employer assigns without consent	X	X	
Employer fails to comply with CDM	X	X	X
Suspension of works for specified period due to:			
specific AIs	X	X	
late instructions	X	X	
employer's labour	X	X	
failure to give access	X	X	
any cause, by the employer			X
Employer becomes insolvent	X	X	X
Employer fails to make premises available			X
By either party			
Suspension of works for a specified period due to:			
force majeure	X	X	
damage due to insured risks	X	X	
civil commotion	X	X	
architect's instructions due to default of statutory authority	X		
hostilities	X		
terrorist activity	X		

in the case of MW 98. One or two points, however, are worth noting. Prior notice of default must be given. The AIs are instructions given in regard to:

- correction of inconsistencies;
- variations;
- postponement of the Works.

In order to be able to determine on the ground of late instructions, the contractor must be able to show that the architect failed to comply with clause 5.4 (JCT 98) or clause 1.7 (IFC 98). 'Employer's labour' refers to suspension of the Works caused by the employer or persons employed directly by him in accordance with clause 29 (JCT 98) or clause 3.11 (IFC 98). Such suspension may be due to the carrying out of such work or failure to carry it out or failure to supply materials which the employer has agreed to supply.

Failure to give access under JCT 98 and IFC 98 is qualified by the phrase 'in due time'. Failure to give access is not the same as failure to give possession but clearly, if the employer does not give access, it may amount to the same thing. Therefore, the employer is expected to give access at such a time that the contractor can take possession of the site on the appointed day. If that day is deferred under clause 23.1.2 or 2.2 of JCT 98 or IFC 98 respectively (see section 6.1), the 'due time' becomes the new date for possession. It can be appreciated that, if the employer defers possession for the usual maximum of six weeks and then fails to give access, it may be a total of 12 weeks before the contractor, after a further 14 days' notice, can determine under this ground. In practice, the employer pays a heavy penalty in terms of extensions of time and loss and/or expense for deferment of possession and further suspension due to lack of access may well entitle the contractor to common law remedies.

Suspension by the employer under MW 98, for any reason, gives grounds for determination if it continues for one month continuously. Since that contract does not, in any event, allow the employer to suspend the Works for even one day, his suspension would be a breach of contract at common law which would normally allow the contractor to treat the contract as at an end. The inclusion of this ground is, therefore, something of a mystery. If the employer suspends the Works under this contract, the contractor should reach for his legal advisor without delay.

If the employer becomes insolvent and becomes bankrupt or makes an arrangement with creditors or, being a company, starts liquidation or a number of other matters associated with insolvency, the contractor may determine his employment.

Determination by either party

There are a number of grounds in JCT 98 and IFC 98 enabling either employer or contractor to determine. Damage to Works or existing property has been dealt with.

Under JCT 98 clause 28A and IFC 98 clause 7.13 either party may determine if the Works are suspended for three months by force majeure, damage to the Works by one or more of the insured risks or civil commotion. Both force majeure and civil commotion were mentioned in section 6.1 in relation to claims for extension of time. This clause deals with events which are outside the control of either party in a way which is fair to both parties. Three other events have been added to JCT 98 and to IFC 98. They are:

- architect instructions issued in regard to the correction of discrepancies, variations or postponement of work, provided that they resulted from negligence or default of a local authority or a statutory undertaker;
- hostilities involving the UK;
- terrorist activity.

A suspension period of one month only is suggested for these events.

Consequences

The consequences of determination vary depending upon the party carrying out the determination and the clause under which the determination takes place. If the employer determines under JCT 98 clause 27 or IFC 98 clauses 7.2–7.5, the rights and duties of the parties may be summarised as follows:

- The contractor must give up possession of the site. After determination, the contractor no longer has a licence to remain on site and, if he does so, he is trespassing. This obligation is not expressly stated in JCT 98, but it is implied. It was thought that if the contractor disputed determination under JCT 98, he had the right to stay in possession until the determination was shown to be valid [17]. Currently, the contrary and more sensible view prevails that the contractor must give up possession pending arbitration and if the determination is found to be invalid, the contractor will have an adequate remedy in damages. If the contractor refuses to give up possession of the site, it is thought that an English court would now grant an injunction to force the contractor to give up possession [18]. IFC 98 clearly states that the procedures after determination are without prejudice to adjudication or other proceedings disputing the determination. Thus, the contractor must leave site and await the results of any proceedings.
- If the architect so instructs, the contractor must remove his temporary buildings, plant, equipment, goods and materials and ensure that other owners of equipment do the same. If the contractor fails to comply within a reasonable time, the employer may sell such property of the contractor, but not of others, and hold the proceeds to the contractor's credit.
- The employer may employ and pay others to complete the Works.
- Unless the determination is due to insolvency, the contractor must

assign all benefits in agreements with sub-contractors and suppliers to the employer, if the architect so instructs within 14 days of the date of determination. The employer is entitled to pay these firms and deduct such payments from money due to the contractor.

- The employer may use any of the contractor's site equipment to complete the Works but not equipment belonging to others without permission.
- The employer need make no further payments to the contractor, even on certificates already issued, until the Works are complete and defects rectified, when the architect must draw up an account stating:

  – the amount of loss and/or expense caused to the employer by the determination, including the cost of having the Works completed by others;
  – the amount already paid to the contractor;
  – the amount which would have been payable for the Works in accordance with the contract.

  The result may be a sum payable to the contractor, but it is more likely to be a sum due to the employer. In effect, the employer is entitled to the difference in cost between the original contract and the actual cost. Proper allowance must be made for variations, professional fees and any other costs incurred by the employer.

In a judgment it was held that 'completion' as referred to in this clause is the same as 'practical completion' [19]. The effect was that the architect must immediately draw up his account when practical completion has been certified. Clearly, it would not be possible for the architect to allow for the additional costs of employing another contractor. In practical terms, it meant that if the employer determined the first contractor's employment two months before practical completion, he was entitled to engage another contractor to complete the Works, but as soon as the Works had reached practical completion, the first contractor must be paid the amount due. After the architect had produced a final account in favour of the second contractor, which could be 12 months or so later, the employer could take action at common law to recover from the contractor whatever damages he had suffered. That, of course, was good news for the contractor and bad news for the employer who was placed in the position of having to pay first and chase the money later. This has been (rather verbosely) corrected. Power has also been introduced for the employer to opt not to continue with the work at all, in which case the contractor is entitled to payment as noted above. To prevent abuse of this power, if the employer does not opt to continue or abandon the Works within six months of the date of determination, the contractor may serve notice on the employer requiring a decision.

The employer has the option of not completing the Works. He must notify the contractor within six months of determination. He must follow the notice with a statement setting out the value of work properly carried out together with any other monies due to the contractor under the contract and any loss and expense due to the employer arising from the determination. Taking into account amounts previously paid, the balance must be paid to either employer or contractor as before. If the employer has not acted by the end of the 6 months period, the contractor is entitled to require in writing that the employer must state whether he will or will not complete the work (clause 27.7.2).

If the contractor determines under JCT 98 clause 28 or IFC 98 clauses 7.9 and 7.10, the rights and duties of the parties may be summarised as follows:

- The contractor must remove from site all his temporary buildings, plant, equipment, goods and materials as soon as reasonably possible after determination. In doing so he must take precautions to prevent damage, injury or death in respect of persons and property for which he was liable to indemnify the employer under the insurance clauses of the contract. Thus, the contractor must not leave the Works in a dangerous condition. He must give his sub-contractors similar facilities to remove their property.
- The contractor is entitled to receive, within 28 days, the whole of the retention, but subject to any right of deduction which the employer may have had *before the date of determination*.
- The contractor is entitled to be paid, without waiting for completion of the Works:
  - the total value of the Works at determination;
  - any sum ascertained under the loss and/or expense clause;
  - reasonable cost of removal of his property from site;
  - the cost of materials properly ordered for the Works for which the contractor has paid or is legally bound to pay;
  - any direct loss and/or damage caused to the contractor by the determination after taking into account amounts already paid to the contractor under the contract.

The sting in the tail for the employer is the direct loss and/or damage to which the contractor is entitled. Among the damages to be included is the loss of profit which the contractor would have received had the contract proceeded to completion [20]. If the determination occurs after only a few weeks of a long contract, the employer will find this item particularly expensive. Of course, the contractor must demonstrate that, if the contract had continued, it is likely that he would have made a profit. In recent years this might prove a problem to many contractors.

Where determination occurs by notice from either party under JCT 98 clauses 22C or 28A, or under IFC 98 clauses 6.3C.4 or 7.8.1, the rights and duties of the parties are the same as if the contractor had determined under clauses 28 (JCT 98) or 7.5–7.6 (IFC 98) except that the contractor has no right to any loss and/or damage arising from the determination. JCT 98 provides that the contractor has a right to loss and/or damage if the event concerns damage to the Works by an insured risk due to the employer's negligence.

The consequences of determination under MW 98 are very brief. If the employer determines, the contractor must give up possession of the site and the employer need make no further payment until completion of the Works and the making-good of defects. If the contractor determines, the employer must pay a fair and reasonable sum for the work done and the cost of the contractor's removal of his property. Neither party has the right under this contract to claim loss and/or damage caused by the determination and the employer is not entitled to make use of the contractor's plant on site to finish the Works. Common law rights are preserved, but in order to obtain damages for loss of profit, the contractor would have to demonstrate to the arbitrator or to the court that the employer's breach was such as would entitle the contractor to treat the contract as repudiated.

In conclusion, the following basic points should be noted:

- MW 98 determination provisions do not provide for damages.
- A party contemplating determination should take advice. There is a grave danger that a party trying to determine will thereby commit a breach entitling the other party to determine instead [21].
- From the employer's point of view, determination is always more expensive – in time and money – than persevering to the completion of the contract with the original contractor. This is true no matter who determines.
- Determination should not be lightly threatened. In more ways than one it is the last resort.

References

1. W Nevill (Sunblest) Ltd v Wm Press & Son Ltd (1981) 20 BLR 78.
2. J Jarvis v Westminster Corporation [1969] 3 All ER 1025 CA.
3. Token Construction v Charlton Estates (1973) 1 BLR 48.
4. William Tomkinson and Sons Ltd v The Parochial Church Council of St Michael (1990) 6 Const LJ 319.
5. Simplex Concrete Piles L v Borough of St Pancras (1958) 14 BLR 80.
6. William Tomkinson and Sons Ltd v The Parochial Church Council of St Michael (1990) 6 Const LJ 319.
7. Pearce & High v John P Baxter & Mrs A Baxter (1999) BLR 101.
8. English Industrial Estates Corporation v George Wimpey & Co Ltd (1972) 7 BLR 122.

9. Davis Contractors Ltd v Fareham UDC [1956] 2 All ER 145.
10. Thomas Feather & Co (Bradford) Ltd v Keighley Corporation (1953) 52 LGR 30.
11. J M Hill & Son Ltd v London Borough of Camden (1981) 18 BLR 31.
12. Greater London Council v Cleveland Bridge & Engineering Co Ltd (1986) 8 Con LR 30.
13. West Faulkner Associates v The London Borough of Newham (1995) 11 Const LJ 157. See also p. 9.
14. John Jarvis Ltd v Rockdale Housing Association Ltd (1985) 5 Con LR 118.
15. D R Bradley (Cable Jointing) Ltd v Jefco Mechanical Services (1988) 6-CLD-07–1.
16. Nash Dredging Ltd v Kestrel Marine Ltd [1986] SLT 62.
17. Hounslow London Borough Council v Twickenham Garden Developments L (1970) 3 All ER 326.
18. Kong Wah Housing v Desplan Construction (1991) 2 CLJ 117. Chermar Productions v Prestest (1991) 7 BCL 46.
19. Emson Eastern L v E.M.E. Developments (1991) 55 BLR 114.
20. Wraight Ltd v P H & T (Holdings) Ltd (1968) 8 BLR 22.
21. Lubenham Fidelities & Investments Co Ltd v South Pembrokeshire District Council and Wigley Fox Partnership (1986) 6 Con LR 85.

8 Dispute resolution

8.1 Adjudication

The Housing Grants, Construction and Regeneration Act 1996 section 108, gives the right to any party to a construction contract to submit disputes to an independent adjudicator. In 2000 it is already clear that the system is becoming very popular and it is expected that most disputes will be resolved in this way. Section 108 also provides that the contract must:

- enable a party to give notice of adjudication at any time;
- provide a timetable to secure an appointment and referral to the adjudicator within seven days of the initial notice;
- require the adjudicator to make a decision in 28 days from referral;
- allow the adjudicator to extend the period as agreed by both parties or for 14 days if just the party who started the procedure agrees;
- impose a duty on the adjudicator to act impartially;
- enable the adjudicator to investigate the facts and relevant law without being requested to do so.

The adjudicator's decision will be binding until otherwise agreed by the parties or until the matter is referred to arbitration or litigation (if there is no arbitration agreement). Neither the adjudicator nor his employees or agents will be liable for actions or omissions done as part of the adjudication.

As and when the referral to adjudication is made, both it and any accompanying documents that are sent with it to the adjudicator must simultaneously be copied to the other party. Whereas no particular form or content is dictated for the notice to refer, the same cannot be said of the referral itself which must clearly set out:

- particulars of the dispute or difference; *and*
- a summary statement of the contentions relied upon; *and*
- a statement of what remedy or relief is being sought.

The contract gives the adjudicator wide powers. He may set his own agenda and discretion to take the initiative in ascertaining the facts and the law in relation to the matter(s) referred to him. Moreover, he may:

- apply his own knowledge and/or expertise;
- open up, reviewing or revising any opinions, notices decisions and the like previously given under the contract;
- visit the site or any other relevant premises used for the preparation of work in connection with contract and/or call on the parties to carry out testing or opening-up of work, or further testing or further opening-up;
- take technical or legal advice and/or, after notice to that effect to the parties, make enquiries of the parties' employees or other representatives.

Once the referral is made then the non-referring party has right of reply. It should be noted that a contractual right of reply is not a strict requirement of the Act. The Act requires merely that the referral be made within seven days of notice to refer and says nothing about the non-referring party's rights of reply, let alone when that reply should be given or what it should contain. Without doubt a right of reply must exist and each of the contracts under consideration give seven days from the date of referral for the response to be made if desired.

Unless and until agreement is reached to be bound or until the adjudicator's decision is ratified or reversed in arbitration or litigation, the parties are bound by it. They must give effect to it and if they fail to do so legal proceedings may be begun in order to secure such compliance, even where all disputes and differences are agreed under the contract to be referred to arbitration as opposed to litigation. It is clear that the courts have a policy of enforcing adjudicators' decisions whether they are good or bad [1].

It is worthy of note that, subject only to any act of bad faith either by the adjudicator or by anyone acting as his employee or agent, the adjudicator and such employees or agents enjoy immunity from liability for anything done or not done in the discharge or purported discharge of his functions as adjudicator. The adjudication provisions are to be found in clause 41A of JCT 98, clause 9A of IFC 98 and supplemental condition D of MW 98.

8.2 Arbitration

Arbitration is the time-honoured system of settling disputes. In essence it is when two parties who are in dispute about something agree between themselves to ask an independent third party to settle the matter between them and they agree to abide by his decision.

Arbitration as understood in regard to building disputes has tended to become somewhat cumbersome and peopled by large numbers of solicitors, barristers and expert witnesses. In serious cases involving very large amounts of money it is perhaps understandable why this should be so. It does not seem to be generally appreciated, however, that arbitration can be a relatively quick process if the dispute is minor and dealt with on the basis of documents only. Both parties can agree that solicitors will not be involved and a settlement can be achieved with the minimum of fuss.

The main advantage of arbitration over litigation in the courts is privacy. No one but those immediately involved need know the details of the dispute and the arbitrator's decision (known as his award). Another advantage is that the parties can choose an arbitrator with technical knowledge who understands, better than a judge, the day-to-day problems of the construction industry. Arbitration can be quicker than litigation provided both parties are anxious to proceed. If there is to be a hearing, it can be arranged in a place and at a time to suit both parties.

The disadvantage of arbitration is that the arbitrator, unlike a judge, has very limited powers to move events on if one of the parties is determined to go slow. Arbitration can become very expensive.

Arbitration is governed by the Arbitration Act 1996. Where two parties have entered into a contract which provides that disputes are to be settled by arbitration it effectively prevents one party from deciding to pursue the dispute through the courts. The other party will be granted a stay of proceedings to allow arbitration to take place. The courts used to have a discretion under certain circumstances to allow litigation to proceed, but under the 1996 Act there is no exercise of discretion and the only ground to allow litigation to continue and refuse a stay would be if the arbitration agreement was void or inoperative.

It is unusual for an Act of Parliament to set out principles but, importantly, this Act sets out the underlying principles in section 1. These principles are the basis under which the Act will be interpreted. The principles are as follows:

- The object is to obtain the fair resolution of disputes by an impartial tribunal without unnecessary delay or expense.
- The parties should be free to agree how their disputes are resolved subject only to the public interest.
- The Courts should not intervene except as provided by the Act.

One important point stems from this. Many of the provisions are non-mandatory which means that the parties can exclude them if they so agree. Note that any such agreements must be in writing to be effective. It is vitally important that parties decide carefully which of the non-mandatory provisions, if any, are not going to be adopted. The following are some of the more interesting points:

- The arbitrator must act fairly and impartially.
- The arbitrator must decide how the arbitration is to be run, including the manner of submissions and evidence. If the arbitrator thinks fit he can adopt an inquisitorial role and take the lead in asking questions to find out the facts.
- The arbitrator may order inspection, photography, custody, sampling or experiments on property owned or in the possession of either party.
- The Act now puts a positive obligation on the parties to pursue the claim without delay and to attend hearings.
- The arbitrator can now order the parties to do things and order them to perform their side of a contract.
- The arbitrator can now award simple or compound interest as he sees fit.
- The arbitrator can now 'cap' the costs that can be recovered in an arbitration. This should dissuade anyone who is intent on going to arbitration to settle a score rather than genuinely to settle a dispute. This power may encourage parties to exercise restraint and avoid unnecessary expense.

Although agreements to arbitrate must still be in writing, 'agreement in writing' is given a very broad interpretation so that virtually any form of record, whether signed or not, will qualify.

If both parties have signed a standard form contract containing an arbitration clause, there is little doubt that they have entered into a binding agreement to arbitrate; but what is the position if one party has simply written to the other confirming agreement to enter into a contract on standard terms (e.g. JCT 98)? An arbitration clause is really a separate contract existing alongside the main contract and capable of existing after the main contract has ended. Therefore, if it is intended to incorporate the arbitration clause it is important to refer to it in clear words [2]. Failure to do so, however, is not necessarily fatal [3].

There is an arbitration clause in JCT 98 article 5 and clause 41B, in IFC 98 article 9A and clause 9B, and in MW 98 article 7A and supplemental condition E. Arbitration, therefore, is the principal binding method of settling disputes in these contracts (but see section 8.3). There are some differences in the three arbitration provisions and JCT 98 will be considered first.

JCT 98

The most important point is that a dispute or difference must have arisen between the parties, i.e. the employer and contractor. The architect is not a party to the contract and, therefore, he cannot be a party in the arbitration proceedings although he can be a witness. The article specifically makes reference to disputes with the architect on the employer's behalf. When exactly a dispute has arisen will be something which should be obvious but, basically, the two parties must be in disagreement [4].

The subject matter of the dispute can be the interpretation of contract provisions or any other matter arising in connection with the contract, including items left to the architect's discretion; withholding of certificates, adjustment of the contract sum, unreasonable withholding of consent or rights and liabilities of the parties; in fact, virtually any kind of dispute provided it bears a relation to the contract [5].

The party seeking arbitration must write and give the other party a notice of reference to arbitration. If they fail to agree on a mutually acceptable person within 14 days of the initial request, either party may write to whomever is named in the Appendix as the appointor. The appointor will be either the President or Vice-President of the Royal Institute of British Architects (RIBA), of the Royal Institution of Chartered Surveyors (RICS), or of the Chartered Institute of Arbitrators (CIArb). If it becomes necessary to take this latter course, there is a form to be completed and a fee to be paid. It is worth noting that an appointment made in this way, which is binding on the parties and the person appointed, is almost impossible to remove.

There is a term in the clause which attempts to ensure that if the employer or contractor is already involved in arbitration with a nominated sub-contractor or supplier about issues which are related to the dispute that has arisen between employer and contractor, then the dispute under the main contract will be referred to the arbitrator already appointed. This is a sensible provision to try to save time and expense and to enable the arbitrator to apportion damages and costs and generally to make his award in the same way as a judge in the High Court where there is a procedure for joining defendants and third parties. To be fully effective, the appropriate sub-contracts and agreements between employers and nominated sub-contractors must contain provisions which dovetail into the main contract provisions. Forms NSC/W and NSC/C contain such provisions. If either the employer or the contractor considers that the arbitrator already appointed to decide, say, the sub-contract dispute is not qualified to decide on the main contract issue, either may require the dispute to be referred to a different arbitrator.

The arbitrator has wide powers under the Arbitration Act and the contract gives him additional powers to direct measurements and valuations, to ascertain and award any sum which should have been included in a certificate to open up, review and revise any certificate, opinion, decision, requirement or notice, and to determine all matters in dispute as if the certificates, etc. had not been given. Some important amendments were made to the arbitration clause in 1988. The arbitrator was given additional power to rectify the contract so that it accurately reflects what the parties agreed. This enables the arbitrator to correct errors and certain specific kinds of mistake. It has been held that he always had this power. The new provision simply recognises the point expressly.

The arbitration is to be conducted under the Construction Industry Model Arbitration Rules (CIMAR). These are set out in full in a handy booklet (available from the Joint Contracts Tribunal). An important feature is a choice of procedure. The procedure extends to the taking-up of the award. The choice of procedure is welcome: documents only, full procedure or the short procedure with a hearing, which may be quite informal and may even be held on site. It is not open to the parties to agree to dispense with the CIMAR.

The award of the arbitrator is stated to be final and binding. In practice, arbitration awards do tend to be final and binding because, although there is provision in the Arbitration Act 1996 for an appeal to be made, it is surrounded by conditions. The appeal must relate to a question of law arising from the award. There is no appeal from facts decided by the arbitrator. All the parties must consent to the appeal or the High Court must give permission. The High Court will not give permission unless it thinks the question of law involved will substantially affect the rights of one of the parties.

The arbitrator is immune from actions of negligence although he may be removed and his award overturned in certain limited circumstances. It used to be the custom for arbitrators to give awards without reasons, but the 1979 Act enabled the parties to request reasons or the courts to order them to be given. It appears to be largely due to the determination of the courts that reasoned awards have not led to a spate of appeals.

The courts are reluctant to give permission to appeal to the High Court and appeals to the Court of Appeal are virtually impossible unless the matter involves something of public importance [6]. The parties state that they agree and consent that either of them may apply to the courts to decide any question of law arising in the course of the arbitration or appeal to the High Court on a question of law arising from the award or apply to the High Court for a decision on a point of law which arises during the arbitration. This appears to overcome the necessity of seeking permission from the court itself [7].

English law is to be used in deciding any point under the contract (clause 1.10). This is the case irrespective of the nationality or homes of the parties or the situation of the Works. It is important to settle this point because, in the case of an arbitration between two parties of differing nationalities over a contract in a third country, the position could become complicated. It is always open to the parties to agree that a different system of law is to be applied if they so wish and a suitable amendment must be made in that case.

IFC 98

The arbitration clause in IFC 98 is almost identical to JCT 98. The contract gives the arbitrator the same powers as under JCT 98. Provision is made for a third party joining procedure.

MW 98

The arbitration provision in MW 98 is similar to that in IFC 98. Either the RIBA, RICS or CIArb may be the appointing body in the event that no agreement can be reached.

Reference to arbitration may be made at any time. In view of the relatively short contract periods envisaged by most contracts let on MW 98 terms, arbitrations are likely to take place after practical completion. The 1998 edition of the CIMAR are to be used.

8.3 Litigation

Each contract now has provision for litigation instead of arbitration. It used to be thought that the arbitrator's powers under the contract were wider than that of the court. For example, the arbitrator had the power to open up and review the architect's decisions. That view was largely based on a case [8] which has recently been overruled in the House of Lords [9]. To bring litigation into effect, it is necessary to delete the arbitration article and clause in the particular contract.

8.4 Points to note

The arbitration clause in a contract will survive the contract itself to allow the arbitrator, for example, to decide whether the contract has in fact come to an end and to determine the consequences for the parties [10].

The majority of references to arbitration are settled or fade away before being determined by the arbitrator. If the reference is not to be decided on the basis of documents only, a hearing will be involved. The parties will have exchanged pleadings, discovery of documents will have taken place and expert witnesses will have reported. The whole process tends to be very expensive.

The moral seems to be: do not resort to arbitration unless there really is no alternative. If arbitration is the answer, keep it as simple as possible and when the award arrives it will be final and binding. At that point, the losing party invariably wonders why he did not agree to settle long before and avoid the costs and mental trauma.

Alternative Dispute Resolution (ADR) is commonly referred to as the modern method of resolving disputes. There is no reference to it in any of the three forms under discussion because ADR is essentially a dispute resolution technique which depends upon the willingness – indeed, the determination – of the parties to settle the dispute. If both parties wish to use ADR techniques, there is nothing to prevent their doing so and the courts actively require their use.

References

1. Macob Civil Engineering v Morrison Construction (1999).
2. Aughton Ltd v M F Kent Ltd (1992) 9 CLD-08–10.
3. Roche Products Ltd v Freeman Process Systems (1996) CILL 1171.
4. Hayter v Nelson and Others (1990) The Times 29 March 1990.
5. Ashville Investments Ltd v Elmer Contractors Ltd (1987) 37 BLR 55.
6. BTP Tioxide Ltd v Pioneer Shipping Ltd and Armada Marine SA. The Nema [1981] 2 All ER 1030.
7. Vascroft (Contractors) Ltd v Seeboard plc (1996) CILL 1127.
8. Northern Regional Health Authority v Derek Crouch Construction Co Ltd (1984) 26 BLR 1.
9. Beaufort Developments (NI) Ltd v Gilbert Ash NI Ltd (1998) 88 BLR 1.
10. Crestar Ltd v Michael John Carr and Joy Carr (1987) 37 BLR 113.

Clause index

JCT 98

IFC 98

MW 98

Subject index